MicroStation V8 for AutoCAD Users

Jeanne Aarhus

THOMSON

DELMAR LEARNING

Australia, Canada, Mexico, Singapore, Spain, United Kingdom, United States

THOMSON

DELMAR LEARNING

MICROSTATION V8 FOR AUTOCAD USERS

Jeanne Aarhus

Vice President, Technology and Trades:
Dave Garza

Director of Learning Solutions:
Sandy Clark

Senior Acquisitions Editor:
James Gish

Senior Project Managers:
Eunice Yeates and Tricia Coia

Senior Content Project Manager:
Stacy Masucci

Marketing Manager:
Deborah S. Yarnell

Academic Marketing Manager:
Gayathri Baskaran

Production Director:
Patty Stephan

Technology Project Manager:
Kevin Smith

Cover Design:
Cammi Mosiman

NOTICE TO THE READER

Publisher does not warrant or guarantee any of the products described herein or perform any independent analysis in connection with any of the product information contained herein. Publisher does not assume, and expressly disclaims, any obligation to obtain and include information other than that provided to it by the manufacturer.

The reader is expressly warned to consider and adopt all safety precautions that might be indicated by the activities herein and to avoid all potential hazards. By following the instructions contained herein, the reader willingly assumes all risks in connection with such instructions.

The publisher makes no representation or warranties of any kind, including but not limited to, the warranties of fitness for particular purpose or merchantability, nor are any such representations implied with respect to the material set forth herein, and the publisher takes no responsibility with respect to such material. The publisher shall not be liable for any special, consequential, or exemplary damages resulting, in whole or part, from the reader's use of, or reliance upon, this material.

About the Author

Jeanne Aarhus has been involved in computer-aided design for more than 20 years. During this time she has been a designer, trainer, administrator, programmer, and consultant using both Autodesk and Bentley products. She is known for keeping her presentations fast-moving and fun, while providing a thorough understanding of the topic. Her specialty is in providing users with the necessary tools for increasing productivity and getting the job done as efficiently as possible, and she continues to focus on maximizing the user's time and efficiency. Jeanne is a graduate of the University of Nebraska and has been certified in all levels of MicroStation and AutoCAD, which enables her to assist users moving from one CAD environment to another easily and proficiently. She has worked in both private and government sectors, and she currently works as a consultant using both Autodesk and Bentley products.

Acknowledgments

It is easy to overlook the many people who contribute to any technical writing project, but I would like to thank all of the individuals who contributed ideas, unending support, and technical expertise over the years. Without their contributions I could not have written this book. I would specifically like to thank Gino Cortesi, Samir Haque, Melissa Hook, and the many members of the Bentley support staff who answer those endless support questions with professional expertise and a cheerful smile. I would also like to send a special thank you to Ray Bentley for his technical expertise and unlimited patience with my endless "AutoCAD" questions and newsgroup inquiries, which are at times trying at best.

I must also thank my many Autodesk contacts for their support and friendship over the years while I too learned to transition from one CAD product to another. Without their support and expertise I could not have survived the transition all those years ago!

Then there are the folks from Delmar Thomson Learning. Without their help, this book would still be scattered ideas and unorganized thoughts just waiting to be put down on paper. I want to thank Carol Leyba for her patience and dedication to the design and formatting of this book and Daril Bentley for his editing expertise and attention to detail. I also need to thank Eunice Yeates for keeping us all on track and on time, which I confess was not an easy task! Finally, I would like to thank Jan Cella for his assistance, many contributions, and useful suggestions during the initial phases of this book.

I dedicate this book to my husband, Steve, who has provided me with unwavering support and patience over the last 20 years while I traveled the world training users and presenting at conferences. Thank you, PB!

Contents

Introduction

Welcome to the "other side" of CAD. At this point you are probably feeling a little nervous, or maybe extremely nervous, about your decision to learn a new CAD application. Don't panic. Many users have survived this transition and even thrive in the MicroStation environment today. Dual CAD environments are becoming more and more common, and once you have read and completed the exercises in this book you will discover what the differences and similarities are between MicroStation and AutoCAD.

For this book, we are focusing on the current releases of both CAD applications: MicroStation V8 2004 and AutoCAD Release 2006. First and foremost, remember that no two CAD applications are exactly alike. However, as technology progresses (and the "Windows" environment invades just about every aspect of our computer applications), MicroStation and AutoCAD are becoming more and more similar. However, MicroStation and AutoCAD are still different CAD applications in many ways, specifically in design and usage. Your goal is to make this transition as seamless as possible. Learn the differences, understand the limitations, and use your AutoCAD knowledge to learn MicroStation. So, take a deep breath and smile. You are about to embark on an adventure where change is progress. The following are a couple of quotes you might want to keep in mind as you jump in head first and turn this page.

> It is not necessary to change. Survival is not mandatory.
> —W. Edwards Deming

> The way we see the problem is the problem.
> —Stephen Covey

THE HISTORY OF MICROSTATION

You might be wondering when and where MicroStation was developed, so here is a short history lesson on MicroStation. Before there was MicroStation, there was PseudoStation. Before there was PseudoStation, there was IGDS.

In the late 1960s and early 1970s, a company called M&S Computing (renamed Intergraph in 1980) developed a software program for NASA and the Apollo mission programs. This software program later developed into a multipurpose CAD platform called IGDS (Interactive Graphic Design System), which ran on a mainframe computer with proprietary user computers. These proprietary user computers were powerful but expensive. During this time frame the personal computer (PC) was making its introduction into the business world, and many IGDS users began to show interest in running this software on a PC. Intergraph pursued hardware alternatives, but ventured into the UNIX environment instead of the DOS/Windows environment.

In the 1980s, Keith Bentley developed a software product called PseudoStation to allow users to open and work with IGDS files on the PC platform. Joined by his brothers Ray, Barry, and Scott, Keith founded Bentley Systems to further the development of this product. PseudoStation was renamed MicroStation around 1980 and over the years has evolved into a powerful CAD platform and the core product for the many verticals offered today by Bentley Systems.

In the 1980s, Bentley and Intergraph worked closely to develop MicroStation for many platforms, including DOS, Windows, UNIX, and Mac. Today, Bentley Systems continues the development of MicroStation and develops over 200 other software products for civil, architectural, plant and process engineering, and geospatial and mapping applications.

FILE FORMATS AND INTEROPERABILITY

MicroStation uses the DGN file format, which is typically smaller and faster than the DWG file format. The DGN file format had not changed since its inception in the 1980s. We applaud the efforts made by Bentley Systems to ensure a stable and migration-free environment through all of the versions and technology changes seen in the past 25+ years. The first DGN file format change occurred in 1999, with the release of MicroStation V8. This file format change provided considerable enhanced features, and was a departure from the "DGN never changes" concept. More detailed information on the specifics of the DGN file format changes can be found later in this book.

The DWG file format is the most common CAD file format being used by many CAD applications today, including MicroStation. MicroStation V8 introduced the ability to open, save, and edit DWG files with no translations required.

LEARN TO SPEAK THE LANGUAGE

One of the first barriers for AutoCAD users is the language barrier. The following tables demonstrate that a simple conversation can result in a multitude of misunderstandings on *both* sides. Learning to speak the "other" language can minimize this barrier and simplify the entire process.

AUTOCAD TERMS AND THEIR MICROSTATION EQUIVALENTS

AutoCAD	MicroStation
Object or entity	Element
Layer	Level
Property	Attribute
Block or WBlock	Cell
Drawing file or drawing database	Active design file
Xref or reference file	Reference file
Prototype or template file	Seed file
Explode	Drop
Current	Active
Attribute	Tag
Zoom Extents	Fit
Zoom Window	Window Area
N/A *(closest match is the Window/Crossing tools)*	Fence
Hatch	Pattern
2D Solid or Solid Hatch	Solid Fill
Grip	Selection handle
Selection Window	Fence or Selection INSIDE, or Polygon Selection (SE/J *PowerSelector*
Selection Fence	Line Selection (SE/J *PowerSelector*
Selection Crossing	Fence or Selection OVERLAP, or Polygon Selection (SE/J *PowerSelector*
N/A EXCLUDE Window *(Express Tools only)* EXCLUDE Crossing *(Express Tools only)* N/A	Fence or Selection CLIP VOID VOID OVERLAP VOID CLIP
ByLayer or ByBlock	ByLevel or ByCell
Level 0 (zero)	Default level
Models	Modelspace and Paperspace
Sheet	Layout

AUTOCAD ELEMENTS AND THEIR MICROSTATION EQUIVALENTS

AutoCAD (DWG)	MicroStation (DGN)
Line	Line
Point	Line *(zero length)*, DL = 0, point
Polyline or LWPolyline	Open Smartline Line String
Polyline or LWPolyline	Closed Smartline (Shape/Complex Chain)
Text	Text
MText	Text Node

AutoCAD (DWG)	MicroStation (DGN)
N/A (closet match is *Attributes*)	Enter Data Field
Polyline, Face, Polyline Mesh, solid	Shape (< 4 vertices)
Polyline, Polyline Mesh	Shape (> 4 vertices)
Polyline/Line, Hatch	Shape/Hatch
Polyline/Line, Hatch	Shape/Xhatch
Polyline/Hatch	Shape/Pattern
Polyline/Solid Fill Hatch	Shape/Opaque Fill
Block or WBlock	Cell
Block or WBlock	Shared Cell
N/A (DesignCenter, single block stored per file)	Cell Library
Circle	Circle
Ellipse (> R13), Polyline (< R13)	Ellipse
Arc	Arc
Spline	Curve
Spline	Curve Stream
Dimension	Dimension
Ellipse (> R13), Polyline (< R13)	Ellipse
Polyline(s) or Multi-line	Multi-line
Attribute or text	Tag
Ellipse (> R13), Polyline (< R13)	Ellipse
Xref	Reference file (design file)
N/A	Self-Referenced file
Viewport	Reference file (sheet file)
Spline	B-splines
Field	N/A
Table	N/A

TEXT CONVENTIONS

The following text conventions are design elements intended to help you make more efficient and productive use of the book.

NOTE: *Notes highlight important information contained in the text or make noteworthy comments on it.*

TIP: *General Tips provide you with information that will help make your user experience more productive and more convenient and/or save you time.*

AUTOCAD TIP: *AutoCAD Tips highlight important information related specifically to the efficient and productive use of MicroStation in relation to AutoCAD.*

HINT: *Hints are found within exercise material. These provide information and reminders that will keep you on track and help you through the process of completing the exercises.*

1: Surviving the Interface

CHAPTER OBJECTIVES:

- ❏ Learn to use tools efficiently
- ❏ Learn about tool frames versus toolbars
- ❏ Understand the Tool Settings dialog
- ❏ Learn to "familiarize" the interface for AutoCAD users

The objective of this chapter is to overcome those AutoCAD interface habits and modify your behavior to use the MicroStation interface as if you had been doing so for years. Although the interfaces initially appear very different, they are really very similar from a "technical" perspective.

This might well be the most difficult aspect of the change you are making in CAD software. Remember, this change can be both "technical" and "emotional" in nature. Every AutoCAD user should focus on the interface differences from a "technical" perspective rather than from an "emotional" one. This can be a tough goal, in that learning a new interface is for many people as "emotional" as it can get.

DIFFERENT SIMILARITIES?

This chapter will focus on the areas of the interface that most affect the transitioning AutoCAD user. With the introduction of the Windows-like interface, most computer applications resemble each other automatically. However, in MicroStation there are some additional concepts that have been integrated into this Windows-like environment to simplify the complex tools needed. For example, toolbars and fly-outs are similar in every application. The sections that follow describe extra features introduced by Bentley.

MicroStation Manager

The first difference is the appearance of the MicroStation Manager dialog. This dialog is the central navigation tool for design files and other CAD-related resources in the MicroStation environment. The unique portions of this dialog that should be noted are the following (see the following figure).

❑ Workspace area

❑ File History feature

❑ Directory History feature

❑ File type display

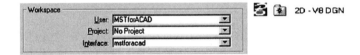

What Is a Workspace?

A workspace is a custom environment that configures MicroStation to a specific discipline, project, task, or individual. Workspaces can help CAD managers and users be more productive by providing a controlled environment for maintaining corporate standards and CAD resources. In general, workspaces consist of configuration files, user interface files, and user preference files.

Configuration files: Contain the overall settings directing MicroStation where to look for resource files, such as those for fonts, cells, designs, and so on. These configuration files are hierarchical in nature and can help CAD managers and users integrate corporate CAD standards and user requirements at varying levels. The level categories available are *System, Site, Application, User,* and *Project.*

User interface files: Allow you to customize toolbars and consolidate the most commonly used tools into a convenient and easy-to-use environment.

User preference files: Store individual user settings to control the "look and feel" of MicroStation. This can include everything from the size and color of dialog boxes and tools to how dialog "focus" is controlled.

The following workspace components are available from the MicroStation Manager dialog.

User settings: Define user-specific configuration settings made to the overall MicroStation setup.

Project settings: Define project-specific configuration settings made to the overall MicroStation setup.

Interface settings: Define user-specific interface settings made to the overall MicroStation setup.

All other components and levels of configuration are defined outside the MicroStation Manager dialog and are generally not modified by the average user.

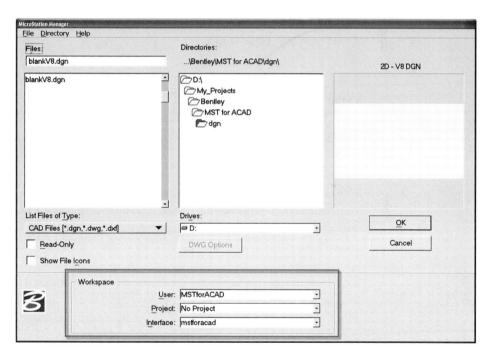

Main Tool Frame

The main toolbar is the central tool chest for MicroStation. This toolbar is unique in that it is actually a tool "frame" that allows the docking capabilities to be controlled in a multi-column fashion. A tool frame also allows for toolbars to be separated (or "torn away") from the main tool frame for quicker access (see following figure).

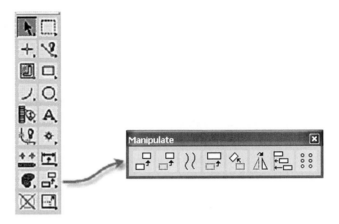

Tear-away Toolbars

MicroStation allows any fly-out toolbar to be "torn away" from a tool frame, providing easy access to commands.

Persistent Commands

All MicroStation commands are persistent in nature. This means that once a command has been selected it remains active until another command has been selected. This is unlike AutoCAD, in which commands end after each use and the user is required to restart the command for consecutive command use.

You can change this behavior in MicroStation via **Workspace > Preferences** and select the *Look and Feel* category. Modify the Single-Click setting to Single-Shot to simulate the AutoCAD single command operation. I recommend you try this new command method before changing this preference—it can save you a lot of picks and clicks.

> **AutoCAD Tip:** *AutoCAD users tend to reselect commands when first learning the MicroStation interface. This is unnecessary and counterproductive because MicroStation commands remain active continuously until a different command is selected.*

In Exercise 1-1, following, you have the opportunity to practice working with the tear-away functionality that allows you to configure your working environment to your liking.

EXERCISE 1-1: USING TOOLBAR TEAR-AWAYS TO CUSTOMIZE THE WORKING ENVIRONMENT

In this exercise you will learn how to arrange toolbars and other collections of commands to customize your working environment.

1 Open the design file *INTERFACE.DGN*.

First, we want to remove the Manipulate toolbar from the Main tool frame and dock it in a more convenient location. This is because these are commands you will typically use several times a day. If you are a Windows user, you probably know that a small black triangle on a button icon means that there are subcommands associated with this button.

2 Select the Copy tool from the main tool frame but hold the left mouse button down long enough to see the "fly-out" toolbar.

3 Drag the cursor into the drawing window far enough to separate and "tear away" this toolbar.

4 Drop the toolbar anywhere in the drawing window. The Manipulate toolbar should now be floating in the middle of your drawing window.

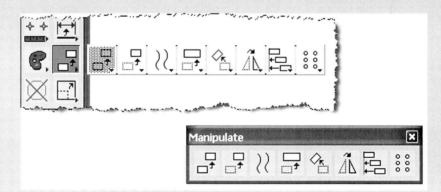

Next, we want to "dock" this toolbar so that it takes up less screen real-estate.

5 Click on the title bar of the Manipulate toolbar. Drag the toolbar to the edge of the application window and note that the "toolbar outline" gets smaller when it is ready to dock. Once you see this change in the toolbar outline, release the mouse button to drop the toolbar in place.

To undock the toolbar, click on the "move bar" and drag the toolbar into the view window.

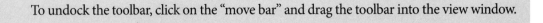

move bar

larger "toolbar outline"

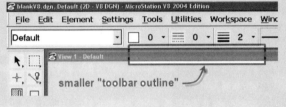

smaller "toolbar outline"

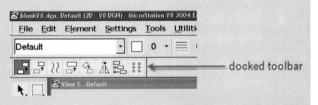

docked toolbar

All toolbars can be "torn away" from the Main tool frame. The only button on the Main tool frame that cannot be torn away is the Delete tool. However, if you look closely you can see that it does not contain the necessary fly-out symbol.

6   Close the file *INTERFACE.DGN*.

Zooming and Panning

The zoom and pan functionality is very similar to that in AutoCAD. Users generally prefer to use a wheel mouse to control the zoom and pan commands.

❑   Roll the wheel mouse away from you (forward) to zoom in, and toward you (backward) to zoom out.

❑   The most common method of panning is to hold down the Shift key while dragging the left mouse button around the drawing. Try to think of your cursor as a car driving around the screen. That should help you get where you want to go. Unfortunately, unlike AutoCAD you cannot use the wheel mouse to pan around in the drawing. However, you can do a different form of panning by rolling the wheel while holding down the Ctrl key to access the Pan Radial option.

❏ Roll the wheel mouse while holding down the Shift key to access the Pan with Zoom option.

Refer to the "Taming the Mouse" section in this chapter for additional settings for wheel mouse control. See Chapter 4 for additional information on view control options.

Command Line

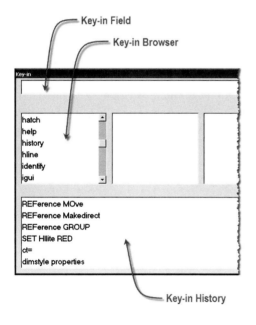

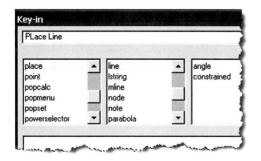

The command line in MicroStation is not displayed by default but can be turned on if you prefer to key in commands. MicroStation commands are generally two- to three-character unique, meaning that you only have to key in two to three characters for MicroStation to figure out what command you want. You can learn these new key-ins using the Key-in Browser dialog.

Select **Help > Key-in Browser** to open the dialog the first time, or you can select **Utilities > Key-in.** Dock the browser dialog horizontally if you plan to use it on a regular basis.

The key-in to place a line in MicroStation is *PLACE LINE*. However, the entire key-in is not required. For example, to key in *PLACE LINE* you are only required to key in the characters *PL* and *L* to make it unique. The remainder of the key-in command is automatically completed.

Refer to the "AutoCAD Key-in Commands" section for additional AutoCAD key-in capabilities.

In Exercise 1-2, following, you have the opportunity to work with the command line.

AutoCAD Key-in Commands

It is common practice for an AutoCAD user to key in commands instead of using the command buttons or pull-down menus. These AutoCAD commands are available as key-ins through the MicroStation environment if needed. In the Key-in browser, start your AutoCAD command with a backward slash (\ ) character. You can use the characters (DWG) instead of the (\ ) character for the AutoCAD key-in prefix. Some configuration changes may be required before these AutoCAD key-ins are available.

Select **Workspace > Configuration** and select the *DWG/DXF* category. Select the *PGP Command Alias file* search path. This search path must be defined to locate the *ACAD.PGP* file with the preferred AutoCAD key-ins.

Select **Workspace > Configuration** and select the *DWG/DXF* category. Select the *DWG Command Prefix* option. This prefix defines the character MicroStation will recognize as the prefix to the subsequent characters representing the AutoCAD command in the PGP file. For example, if the default prefix is a backslash (\ ), the key-in *\e* will run the AutoCAD *ERASE* command, which is then translated to run the MicroStation *DELETE* command.

EXERCISE 1-2: WORKING WITH THE COMMAND LINE

In this exercise you will learn to use the command line with both MicroStation and AutoCAD key-ins.

Using the Command Line with MicroStation Key-ins

1   Open the design file *INTERFACE.DGN*.

First, you need to open the Key-in Browser dialog and dock it at the bottom of the application window in a similar location to AutoCAD's command line.

2   To open the Command Line dialog, go to the pull-down menu **Help > Key-in Browser.** Or you can use the button located on the Primary Tools toolbar.

Before we dock the command line let's review how to use the Browser section (see following figure) to look up key-in commands. For this example, we will look up the *PLACE*

LINE key-in command. To see the Browser section of the dialog, drag the bottom of the command line down until you see additional command "look-up" columns.

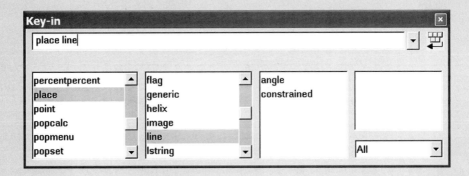

3 In the "first" browser column, select the *PLACE* command.

Once the *PLACE* command is highlighted, the "second" browser column is populated with possible "place" command options. You can teach yourself the majority of MicroStation commands by "browsing" in this dialog.

Once the *LINE* command is highlighted in the "second" column, the "third" browser column is populated with possible "place line" command options.

You do not have to use all command options, as in this case we will not select either *ANGLE* or *CONSTRAINED* as a third command option. We just want to issue the *PLACE LINE* command. You can shorten this key-in using the "unique" portions of the command. The best way to learn these shortcuts is to engage Caps Lock on your keyboard and begin entering the key-in command. When MicroStation figures out what command you are entering, it will automatically fill in the remainder of the key-in. When it does this, you can begin to enter the next command option and repeat the same process.

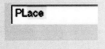

4 Key in the letters *PL* and you should notice that MicroStation completes the remainder of the word *PLace*.

5 Key in the letter *L* and you should notice that MicroStation completes the remainder of the word *Line*.

Now you need to place the Key-in browser in a productive location so that it is available when needed, but out of the way when not needed. Again, AutoCAD users might want to dock this browser at the bottom of the application window. After all, isn't that where you are going to look by default anyway?

6   Dock the Key-in Browser dialog at the bottom of your application window.

Moving Focus

You can move the focus of the application back to the command line using several different methods.

1   While using AccuDraw, the shortcut key combination GK (Go to Keyin) will move the application focus to the Key-in browser automatically. See Chapter 4 for more information on the AccuDraw Utility.

2   You might try using the Esc (Escape) key to move the focus as well. However, this will not work if your user preferences is set to use the Esc key to stop current commands.

3   You can always click in the key-in field to move the application focus manually.

Using the Command Line with AutoCAD Key-ins

Next, let's configure MicroStation to allow you to use your familiar AutoCAD commands directly within MicroStation. See, you don't really have to relearn everything!

1   To access the configuration settings, go to the pull-down menu **Workspace** > **Configuration** and select the *DWG/DXF* category.

2   Highlight the *PGP Command Alias File* option and verify that it is pointed to your *ACAD.PGP* file.

   If AutoCAD is loaded on your computer, this setting will be automatically set during the installation of MicroStation and you can skip to step 4. If it is not pointed to your *ACAD.PGP* file, you must modify this setting manually.

3   Click on the Select button and navigate to the location of your *ACAD.PGP* file.

4   Select the *DWG Command Prefix* option and make note of the "prefix" character defined for AutoCAD key-ins. By default, it should be set to use the backslash (\ ) character.

5 You must exit MicroStation before this modification can be used.

6 Restart MicroStation and open the *INTERFACE.DGN* file.

7 Move the application focus to the command line and key in *\E*. This will run the *DELETE* command in MicroStation, the equivalent of the *ERASE* command in AutoCAD.

Try some of your other AutoCAD favorites and see if they are translated correctly. Pretty cool, right?

8 Close the design file *INTERFACE.DGN*.

Get a Grip on Handles

The handles provided in MicroStation are not the same as AutoCAD's grip functionality. Handles are not associated with any specific commands, but they can be used with any MicroStation command. The most common use of handles would be to modify or to move an element. In Exercise 1-3, following, you have the opportunity to practice using MicroStation's element handles.

EXERCISE 1-3: GETTING A GRIP ON HANDLES

In this exercise you will learn to use MicroStation's element handles. Element handles can be used to make simple modifications. You will use these handles to modify the size of a desk and the location of leader text.

1 Open the design file *HANDLES.DGN*.

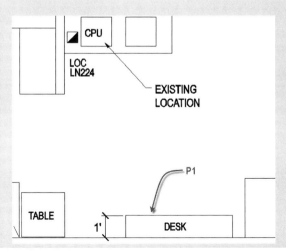

Modify Desk Size

2 Select the Element Selection tool and note that the cursor changes to a selection arrow.

The circle attached to the point of the arrow represents the location tolerance specified in the user preferences area. The smaller the circle the closer your cursor must be to an element for snapping or selection capabilities.

3 Pick the desk element at P1 (on previous page) and note the appearance of element handles.

Pick the desk a second time to reveal additional handles at the midpoints. These handles provide additional modification points for more specific editing requirements.

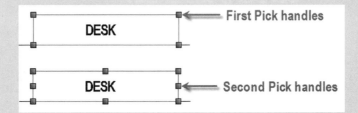

4 Pick the handle at the P2 and note that the handle turns a reddish color when your cursor is over the handle.

Continue to hold down the left mouse button and drag the selected handle to the corner of the adjacent table. This will dynamically change the shape of the desk to match the depth of the table.

An AccuSnap icon will appear when both elements have connected at the vertex. Release the mouse button when you see the AccuSnap icon indicating that a snap point is available.

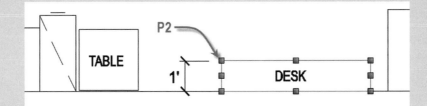

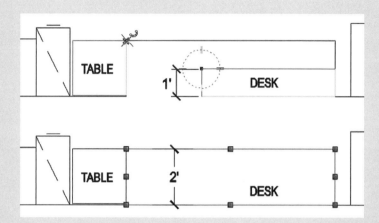

In the next few steps you will learn to use the midpoint handles available when an element is picked twice.

5   Pick the desk at P3 *twice* in order to activate the additional element handles.

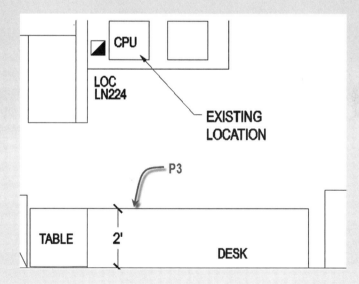

6   Pick the handle at P4 (see figure on next page), continue to hold down the left mouse button, and drag the selected handle straight up using AccuDraw. Key in *2* to modify the depth of the desk by 2 feet.

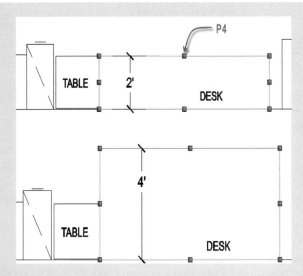

Modify Text Position

In the next few steps you will modify the location of the dimension-note text associated with the desk geometry.

7   Open the design file *OFFICE_ AREA.DGN*.

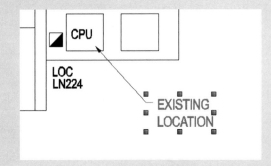

8   Select the text element that reads *EXISTING LOCATION* and note the element handles available. Continue to hold down the left mouse button and drag the text to a new location in the drawing window. Because this text element is a dimension note, the leader line will automatically update to the new text position.

AUTOCAD TIP: *You must select a text element on an actual character. If you miss the characters, the text element will not be selected.*

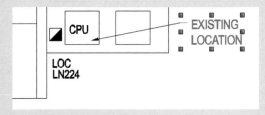

9   Close the design file *HANDLES.DGN*.

Double-click Editing

MicroStation provides double-click editing capability through the use of the Selection tool. Double-click editing is available for text and dimension-text only.

USER PREFERENCES

The following sections walk you through some of the personal user preferences that can ease your transition to the new interface.

Windows Options

MicroStation V8 allows you to establish a Windows-like MicroStation Manager interface environment. To achieve this, select **Workspace > Preferences** and select the *Look and Feel* category. Then activate the *Use Windows File Open Dialogs* option.

The following figure shows a comparison between the default MicroStation Manager and the Windows-like MicroStation Manager.

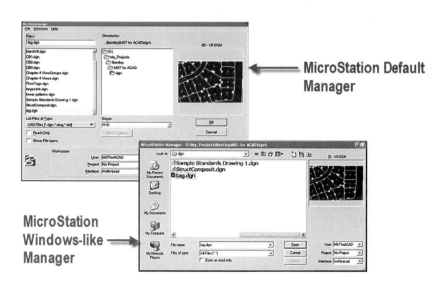

MicroStation Default Manager

MicroStation Windows-like Manager

Esc to Cancel

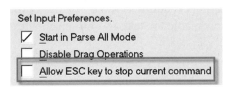

One AutoCAD habit that is extremely difficult to break is the use of the Esc key to end a command. By default, the Esc key does absolutely nothing in MicroStation, but you can enable this feature via the user preferences.

Access this setting via **Workspace** > **Preferences** and select the *Input* category. Activate the *Allow ESC key to stop current command* option to enable cancel command functionality.

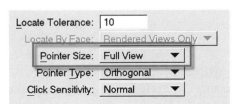

Full-screen Cursor

Another preference to note is the full-screen cursor option still preferred by many AutoCAD users. You can enable a full-screen cursor via **Workspace** > **Preferences** and select the *Input* category. Then select *Pointer Size > Full View*.

AUTOCAD TIP: *In this author's opinion the normal cursor is more functional and easier to use, but as you can clearly note this is a "user" preference.*

In Exercise 1-4, following, you have the opportunity to practice establishing your own user preferences.

EXERCISE 1-4: ESTABLISHING USER PREFERENCES

In this exercise you will learn how to modify the user interface to use your own user preferences. This exercise does not cover all user preference capabilities, but rather those that affect the majority of AutoCAD users.

1   Open the design file *INTERFACE.DGN*.

2   To access user preferences, go to the pull-down menu **Workspace** > **Preferences** and browse the categories available.

These categories affect the appearance of tools, views, and mouse button functions and other global settings.

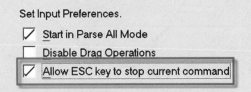

3   Select the *Input* category and activate the *Allow ESC key to stop current command* option. This setting will save you from "breaking" your Esc key during this CAD software transition

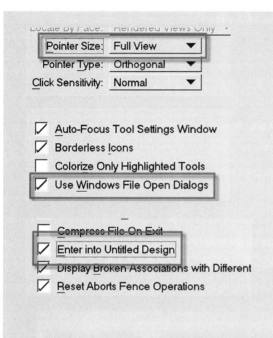

4 Stay in the *Input* category and set the *Pointer Size* option to *Full View*.

5 Select the *Look and Feel* category and activate the *Use Windows File Open Dialogs* option.

This setting will modify the appearance of standard file dialogs to use the Windows-like appearance and functionality.

6 Select the *Operation* category and activate the *Enter into Untitled Design* option.

This setting will bypass MicroStation Manager upon startup and place you directly into a design file similar to the untitled drawing in AutoCAD.

7 Close the design file *INTERFACE.DGN*.

Compress

The *COMPRESS* command, which is similar to the *PURGE* command in AutoCAD, will clean up design files by removing extraneous and unused items. The following options are available for controlling what data is removed from the design file.

Empty Cell Headers: Removes all empty cell headers from the design file. An empty cell header is a cell definition with no graphics associated to it

Empty Text Elements: Removes all empty text elements from the design file. An empty text element is usually a text node with no characters associated.

Text Elements with Spaces Only: Removes any text elements that contain only "spaces" and no other characters. This is usually an "empty" enter data field.

Unused Named Shared Cells: Removes all unused shared cells from the design file. An unnamed shared cell is usually a remnant of previously used shared cells that are no longer being used in the drawing.

Unused Anonymous Shared Cells: Removes all unused shared cells in a design file. Anonymous shared cells are usually a result of Windows clipboard activities, or previous AutoCAD anonymous blocks.

Unused Line Styles: Removes all custom line styles no longer being used in the drawing.

Unused Dimension Styles: Removes all dimension styles no longer being used in the drawing.

Unused Text Styles: Removes all text styles no longer being used in the drawing.

Unused Levels: Removes all levels no longer being used in the drawing.

Unused Nested Attachment Levels: Removes all levels (from reference file attachments) no longer being used in the drawing.

Unused Fonts: Removes all fonts no longer being used in the drawing.

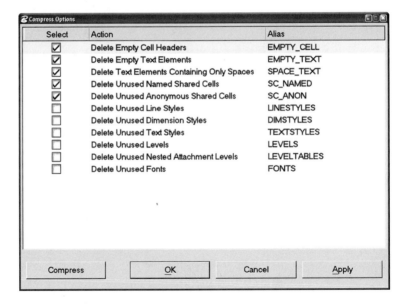

You can access the *COMPRESS* command via any one of the following methods.

❑   **File > Compress > Design**

❑   **Directory > Compress** from the MicroStation Manager dialog

❑Use the Bentley button from the MicroStation Manager dialog to compress the selected file.

UNIQUELY BENTLEY TOOLS

Directory History

Access to directory history is a unique aspect of the Bentley Windows-like environment of MicroStation. This tool provides the user quick and easy access to folder history, which for most users is equivalent to project history. You can access this option from any of the following MicroStation dialogs.

- ❏   MicroStation Manager
- ❏   **File > Open**
- ❏   **File > Close**
- ❏   **File > Save As**

Select Configuration Variable

Another uniquely Bentley option is *Set Configuration Variable* option, accessed through the Directory History button. This option allows the user to select any configuration variable to control directory access. For example, using the configuration variable *MS_BACKUP* will navigate to the *File backups* folder, or *MS_CELL* will navigate to the *Symbol library* folder. These folder locations are defined with MicroStation workspace configuration variables.

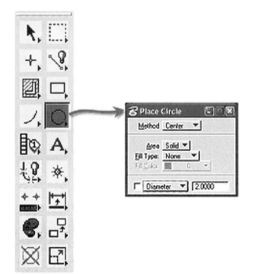

TOOL SETTINGS

Every tool has options specific to its individual usage. These options are displayed in the Tool Settings dialog for easy access and manipulation.

AUTOCAD TIP: *Don't bother closing the Tool Settings dialog, because it will automatically open for every command tool selected. Instead, find a screen location that is the least intrusive to your drawing practices. If this location does not exist, try using the PopSet Utility controls (discussed in material following).*

Tool Settings Dialog Control

The display of the Tool Settings dialog can be controlled by accessing the PopSet utility. When you first begin using the Tool Settings dialog, you either keep turning it off to get it out of the way or constantly move it around the screen trying to find the best place to keep it where it will be out of the way. You soon realize that there really isn't an out-of-the-way place on the screen and the irritation factor sets in. Thus, Bentley has developed the PopSet utility.

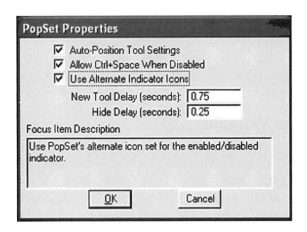

When PopSet is enabled (a green light), the Tool Settings dialog will disappear from view automatically. When a tool is selected, the Tool Settings dialog will reappear next to the selected tool button. When the cursor is moved away from the tool or dialog, the Tool Settings dialog once again disappears automatically.

The speed of this automatic disappearance and reappearance is controlled by PopSet's Properties settings. To access these property settings, right-click on the PopSet button and select Properties. Once PopSet is enabled, there are three ways to force the Tool Settings dialog to reappear.

❑ Select a command or hover the cursor over the previously selected command.

❑ Hover the cursor over the PopSet button.

❑ Press Ctrl + spacebar and the Tool Settings dialog will appear at the current cursor location.

Repeating Commands

MicroStation commands automatically repeat until the next tool is selected.

> **AUTOCAD TIP:** *Remember, you do not have to reselect tools to use them more than once. MicroStation commands remain running continuously until a different tool is selected.*

A Different Class of Command

There are basically two types of commands performed by MicroStation: primary and view.

> **Primary commands:** Include those that draw, modify, and manipulate data in the design file. Use the regular (command) UNDO to reverse your changes.

> **View commands:** Include those that zoom, pan, and "move around" in the design file. Use View Previous and View Next in the View Border to UNDO View Control commands.

How does this affect you? All view commands are "transparent" commands, which means that they will run within primary commands without ending the primary command. Once the view command has been completed, command control is automatically returned to the primary command.

In Exercise 1-5, following, you have the opportunity to practice using the PopSet utility.

EXERCISE 1-5: USING THE POPSET UTILITY

In this exercise you will learn how to use the PopSet utility to maximize your screen real estate and how to work with the Tool Settings dialog efficiently and productively.

Working with the Tool Settings Dialog

1   Open the design file *INTERFACE.DGN*.

2   Select the following tools and note the changes in the Tool Settings dialog. The Tool Settings dialog can drastically change its size based on the active tool. This is another factor that makes it difficult to find a permanent unobtrusive location for this dialog.

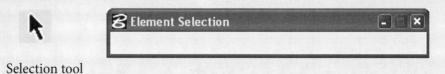

Selection tool

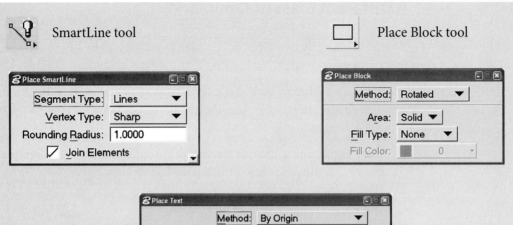

SmartLine tool

Place Block tool

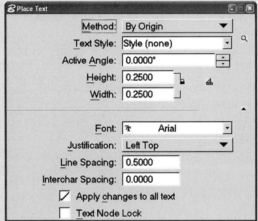

Place Text tool

Activating the PopSet Utility

Now let's activate the PopSet utility so that the Tool Settings dialog can disappear and reappear on demand.

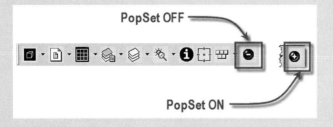

PopSet OFF

PopSet ON

3 Click on the PopSet Utility icon on the Primary Tools toolbar. Note the change in color from red (inactive) to green (active), indicating the utility's status.

4 With PopSet active, select the SmartLine tool but move the cursor away from the SmartLine button and into the drawing view. Note that the Tool Settings dialog disappears once you have moved away from the active button. This "automatic hide" frees up valuable screen real estate.

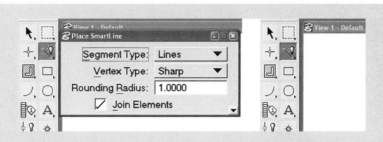

5 You can force the Tool Settings dialog to reappear via any of the following methods.

❑ Move your cursor back to the active tool button.

❑ Move your cursor back to the PopSet Utility button.

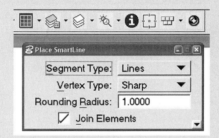

❑ Move your cursor back to the drawing view window and use the shortcut key combination Ctrl + spacebar to get the Tool Settings dialog to appear at your current cursor location.

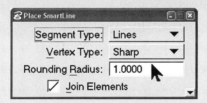

6 Close the design file *INTERFACE.DGN*.

TAMING THE MOUSE

Mouse usage is basically the same in most Windows applications.

Two-button/Wheel Mouse

The following describes the functionality of the components of a two-button mouse with a wheel.

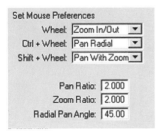

Left button: Selects items and is commonly referred to as the "data" button in MicroStation.

Right button: Stops or resets commands and is commonly referred to as the "reset" button in MicroStation.

Wheel: Zooms in and out.

Additional wheel functionality is available through the User Preferences feature. Select **Workspace > Preferences** and select the *Mouse* category to modify these settings to your preferences.

Two-button Mouse

The following describe the functionality of the components of a two-button mouse.

Left button: Selects items and is commonly referred to as the "data" button in MicroStation.

Right button: Stops or resets commands and is commonly referred to as the "reset" button in MicroStation.

Additional Mouse Features

The following are other mouse features implemented in conjunction with the keyboard.

Ctrl + left button: Adds and removes elements to the current selection set using the Selection tool.

Shift + Ctrl + left button: Selects elements using the Overlap feature of a selection set while using the Selection tool. This is similar to a crossing window in AutoCAD.

AutoCAD Tip: *Use the PAN VIEW command with the Dynamic Display tool setting activated to access a pan command similar to that found in AutoCAD. However, you must use the "data" button (right mouse button) rather than the wheel.*

Precision Snaps

Snapping to drawing elements is required by any user drawing with precision. MicroStation provides two methods for snapping to elements.

AccuSnap: The newest method providing multi-snap capabilities, similar to AutoCAD's object snaps.

Tentative Snap: The original method provided in MicroStation since version 2.0.

AccuSnap

This method is the preferred snap method for most users. AccuSnap provides the user with the ability to combine commonly used snap modes so that more than one is available concurrently. This prevents the user from having to change the snap modes during typical design session operations. This method also provides on-screen feedback to confirm accuracy during drawing and editing.

Tentative Snap

This method dates back to the earlier versions of MicroStation, wherein users typically used a simple two-button mouse. To provide the needed snapping capabilities, the "third" mouse button was simulated by pressing both mouse buttons simultaneously. With the introduction of the three-button mouse, this command moved to the middle button. A wheeled mouse uses the wheel as a third or middle button and simulates all middle button functions.

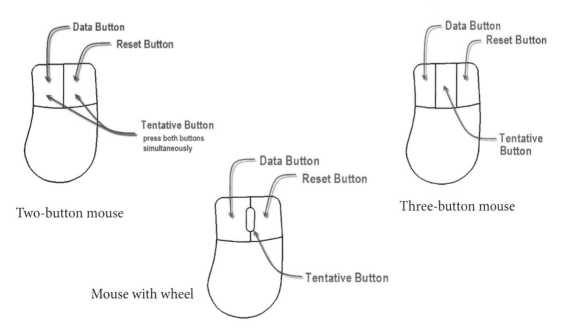

In Exercise 1-6, following, you have the opportunity to practice using the AccuSnap feature and the tentative snap functionality.

EXERCISE 1-6: USING THE ACCUSNAP AND TENTATIVE SNAP FUNCTIONS

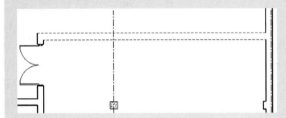

In this exercise you will learn to use two precision snap options provided by MicroStation: AccuSnap and Tentative Snap. Using these options you will add missing wall lines to a design. These are shown as dashed lines in the office layout shown in the figure at left.

AccuSnap Method

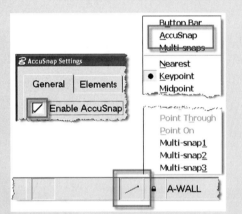

1   Open the design file *ACCUSNAP. DGN*. Before using the AccuSnap method you must verify that AccuSnap is enabled.

To verify that AccuSnap is enabled, go to the status bar located at the bottom of the application window and click on the Active Snap Mode field.

Select AccuSnap from the pop-up menu to open the AccuSnap Settings dialog.

Verify that Enable AccuSnap is activated (toggled On).

Now you are ready to use the AccuSnap method to add the missing wall lines to the design drawing.

2   Select Place Line and move the cursor to the P1 location to identify the start point of the wall (see figure on next page, at top).

When the cursor gets close to the keypoint on the line, the AccuSnap icon (arc with blue dots) should appear, along with a yellow X to identify the snap point.

You must always accept an AccuSnap point, so issue a data point (click the left mouse button) to accept this snap location. Your line should be attached to point P1.

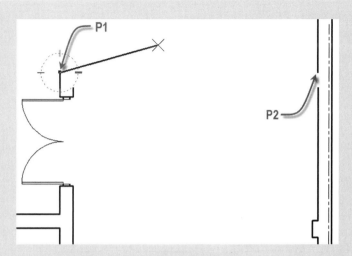

AUTOCAD TIP: *The AccuSnap icon works differently than the OSnap marker in actual operation. The yellow X is the actual marker you need to see before you click on a mouse button. The snap method icon (arc with blue dots) only indicates the type of snap method currently active. The mistake most AutoCAD users make is to click when only the snap method (blue) icon is visible and not the AccuSnap (yellow) icon. Making this mistake causes MicroStation to incorrectly snap to existing geometry.*

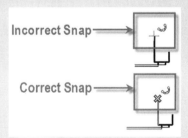

3 Use AccuSnap to snap to P2 and accept it. Use the Reset button (right mouse button) to complete the current line segment.

4 Use AccuSnap to snap to P3 and accept it.

5 Use AccuSnap to snap to P4 and accept it. Use the Reset button to complete the current line segment.

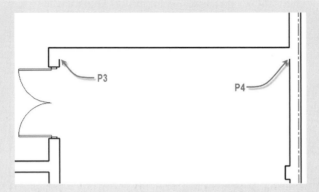

6   Remove both lines using the Delete tool, or by using the Selection tool and pressing the Delete key.

Tentative Snap Method

In the next few steps, you will learn to use the Tentative Snap feature to recreate the same two wall lines. Before using the tentative snap method you must verify the configuration of your mouse buttons.

7   To access the mouse button configuration, go to the pull-down menu **Workspace > Button Assignments**. Verify the Tentative button setting for your current mouse type: two-button mouse, two-button mouse with wheel, or three-button mouse.

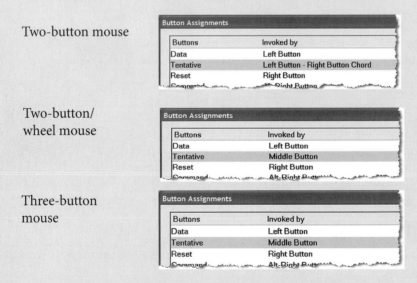

8   If your Tentative setting matches the applicable mouse type, skip to step 10. Otherwise, continue with step 9.

9 Select the Tentative button assignment, move the cursor over the "button definition area," and click on the mouse button you want to use for tentative snap capability. This is generally the middle button or the wheel.

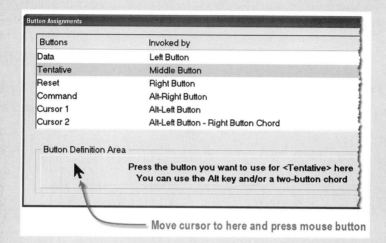

10 Click on the OK button to accept the changes and close the Button Assignments dialog.

Now we are ready to use the tentative snap method to add the missing wall lines to the design drawing.

11 Select Place Line tool and move the cursor to the P1 location to identify the start point of the wall. Rather than using AccuSnap we will tentative snap to the existing wall line.

Issue a tentative snap (configured in the previous step) and the existing line should highlight and a "larger crosshair" should display.

You must always accept a tentative snap, so issue a data point (left-click) to accept this snap location.

Your line should be attached to point P1.

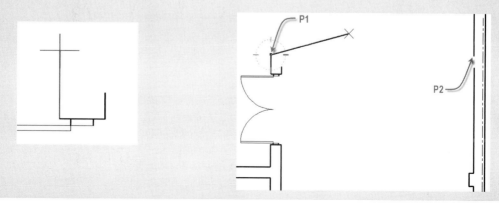

AutoCAD Tip: *If you miss the intended snap point when tentative snapping, the cursor may highlight the wrong element, or display a dashed crosshair. If the wrong element is highlighted, just issue a second tentative snap and MicroStation will continue to cycle through all available snap positions near your cursor. Once the correct element is highlighted, issue a data point (left-click) to accept this location.*

The dashed crosshair indicates that MicroStation could not find an element to snap to. Move your cursor closer and try again.

12   Issue another tentative snap at P2 and accept it. Use the Reset button (right-click) to complete the current line segment.

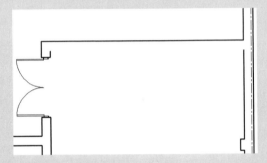

13 Issue a tentative snap at P3 and accept it.

14 Issue a tentative snap at P4 and accept it. Use the Reset button to complete the current line segment.

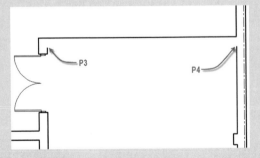

15 Close the design file *ACCUSNAP.DGN*.

POWER DRAFT

All chapter content works the same in PowerDraft.

2: Basic DGN Concepts

CHAPTER OBJECTIVES:

❑   Understand the DGN file format

❑   Learn the drawing setup

❑   Establish standards using seed files

❑   Introduction to the new V8 models

THE V8 DGN FILE

First, let's explore a little history on the V7 DGN file format. The V7 DGN file format has been the foundation of MicroStation since its inception in the 1970s. Even though the software was not called MicroStation yet, the file format was the reliable DGN we all learned to love. This format was very fast and efficient, and more than satisfied the needs of that time frame. With the explosion of Windows and CAD capabilities over the past few years, many of us anxiously awaited an overhaul of the DGN format with the release of V8. The following is a comparison of both DGN file formats.

V7 DGN FILE FORMAT:

❑   Limit of 63 levels based on numbers and names

❑   Limit of 32 MB DGN file size

❑   Limit of 256 K symbol size

❑   Limit of 6 characters for cell names

❑   Limit of 256 reference files

❑   Limit of 101 vertices per element

❑   Limit of 256 characters in a single text element

V8 DGN FILE FORMAT:

❑ Unlimited levels based on numbers and names

❑ Maximum file size of 4 GB

❑ Unlimited symbol size

❑ 500 character cell names

❑ Unlimited references

❑ 5,000 vertices per element

❑ 65,535 characters in a single text element

❑ Unlimited text node size

With these changes, the DGN file has matured into an extremely robust CAD file format. In addition, the V8 DGN file includes the IEEE 754-1985 specification for the storage of coordinate geometry, which means that its drawing plane is about 2 million times larger in the X and Y directions than the V7 DGN file.

There are two types of DGN files: 2D and 3D. Both of these file types define a "global origin," which specifies a unique 0,0 location in the design plane or design cube. This global origin is located in the center of the flat plane and 3D cube.

The 2D DGN file provides a flat 2D drafting environment similar to a sheet of paper. This environment consists of a limited design plane where data can be input using X and Y coordinates only. 2D files are one-third smaller in file size than 3D files.

The 3D DGN file provides a 3D cube drafting environment similar to the AutoCAD model. This 3D environment consists of a limited design cube where data can be input using X, Y, and Z coordinates.

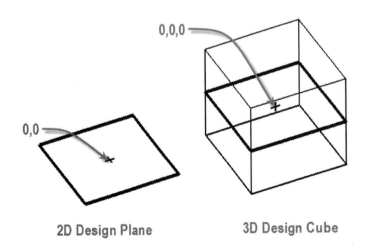

2D Design Plane   **3D Design Cube**

UNITS AND DRAWING SETUP

The DGN file format is controlled by many settings that can be modified by the individual user based on their environment and production requirements. These settings are as follows.

- Active Angle
- Axis Lock
- Coordinate Readout
- Fence
- Isometric
- Rendering
- Stream
- Working Units

- Active Scale
- Color Table
- Element Attributes
- Grid
- Locks
- Snaps
- Views

The most commonly used settings are defined in the sections that follow.

ACTIVE ANGLE

The Active Angle settings define the current angle setting used by many commands, such as place text and place cell. This setting is global to all models in the design file. The active angle can be modified during command operations using the Tool Settings dialog. This setting is saved with the file using the Save Settings command. The Angle Lock setting controls the active angle "round-off" value. Use the key-in *AA=angle value* to set the active angle.

ACTIVE SCALE

The Active Scale settings define the current scale setting used by many commands, such as place cell and scale element. This setting is global to all models in the design file. The active scale can be modified during command operations using the Tool Settings dialog. This setting is saved with the file using the Save Settings command. The Scale Lock setting controls the active scale "round-off" value. Use the key-in *AS=scale value* to set the active scale.

COLOR TABLE

The Color Table settings define the active color table attached to the design file. All 256 colors in the color table can be modified as needed. Color 256 is reserved for the view window background color. Each model in the design file can have a different color table attached. The AutoCAD standard color table (*acad.tbl*) is delivered with MicroStation and can be attached if needed.

This setting is saved with the file using the Save Settings command. You can restore the default color table using the key-in *CT=<space-bar>*.

COORDINATE READOUT

The Coordinate Readout settings define the accuracy of working units and angles displayed during draw and measure commands. This setting is global to all models in the design file. This setting is saved with the file using the Save Settings command.

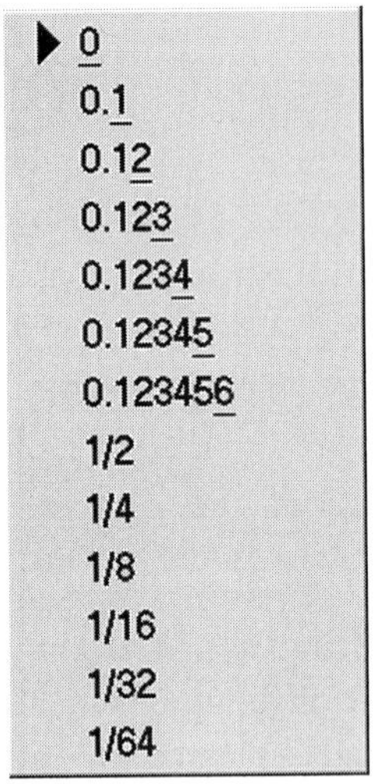

The working unit coordinate options are as follows.

❏   Master units only (MU)

❏   Master units and sub units (MU:SU)

❏   Master units, sub units, and positional units (MU:SU:PU)

The figure at left shows a list of accuracy settings available for working units. The angle unit coordinate options are as follows:

❏   Decimal Degrees

❏   Degrees/Minutes/Seconds

❏   Gradians

❏   Radians

The formats for angular measurement available for working units are as follows.

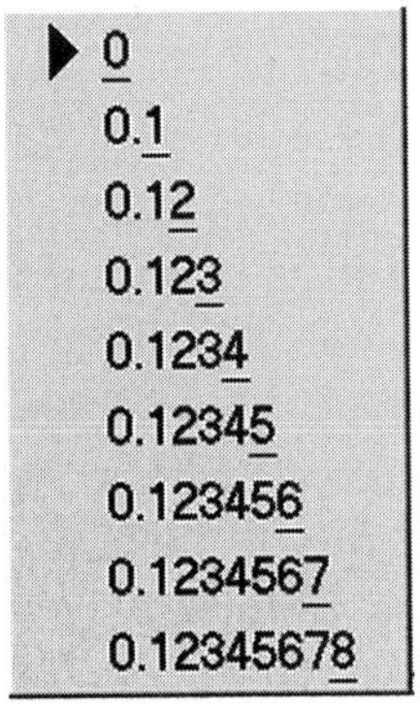

❏   *Conventional:* Angles calculated counterclockwise from a horizontal east base of zero

❏   *Azimuth:* Angles calculated clockwise from vertical north base of zero.

❏   *Bearing:* Angles calculated NE, NW, SE, and SW from a quadrant base of zero

The figure at left shows a list of accuracy settings available for angle configurations.

ELEMENT ATTRIBUTES

The Element Attributes settings define the current settings for level, color, line style, line weight, and element class. Each model in the design file can have different element attributes defined. The element attributes can be modified during command operations using the Attributes toolbar or the Tool Settings dialog. This setting is saved with the file using the Save Settings command.

FENCE

The Fence settings define the current fence mode setting used by fence manipulation commands such as clip reference and copy fence. This setting is global to all models in the design file. The fence mode can be modified during fence command operations using the Tool Settings dialog. This setting is saved with the file using the Save Settings command.

GRID

The Grid settings define the grid unit settings used for drawing elements with a grid guideline. Each model in the design file can have different grids defined. This setting is saved with the file using the Save Settings command. The Grid Lock setting controls whether or not you are forced to use grid points during command operations.

SNAPS

The Snaps settings define the current snap mode used by most commands. This setting is global to all models in the design file. The active snap can be modified during command operations using toolbars, pop-up menus, key-ins, and status bar controls. This setting is saved with the file using the Save Settings command. The Snap Lock setting controls whether or not the Tentative button is available.

VIEWS

The Views settings define the current view window size and viewing options. The view window size can be modified with simple drag operations. These settings are saved with the file using the Save Settings command. The Proportional Resize option can help with specific view window sizing. The pixel size does not include the scroll bars in each view window and can guarantee specific screen capture sizes for publishing and Internet image processing.

WORKING UNITS

Working units define the breakdown of the design plane or design cube into "real-world" units of measurement. These units measurements are based on the meter and conversions to other units of measurement are calculated using 14 decimal places of accuracy. These units can be defined as Imperial or Metric, and you can even define your own custom unit definitions. Modifications to working units no longer affect the size of previously constructed graphics unless you change the positional resolution (advanced) settings. The labels settings defined here control the labels used for dimension elements. These settings are saved with the file using the Save Settings command.

The core items of the unit definition are master units, sub units, and positional units, sometimes referred to in the documentation as MU:SU:PU. Each of these core items can be defined in your units of measurement.

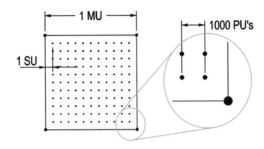

Table 2-1 outlines examples of unit definitions.

TABLE 2-1: UNIT DEFINITIONS

System	Units	Unit Definition
Imperial Architectural	MU:SU:PU	
	1:12:1000	1 foot = 12 inches
		1 inch = 1,000 positional units
		1 foot = 12,000 positional units
	1:12:8000	1 foot = 12 inches
		1 inch = 8,000 positional units
		1 foot = 96,000 positional units
Imperial Civil	MU:SU:PU	
	1:10:100	1 foot = 10 (tenths of a foot)
		1 th = 100 positional units
		1 foot = 1,000 positional units
Metric Architectural	MU:SU:PU	
	1:10:100	1 mm = 10 (tenths of a foot)
		1 mm = 100 positional units
		1 mm = 1,000 positional units
Metric Civil	MU:SU:PU	
	1:10:1000	1 m = 10 mm
		1 mm = 1000 positional units
		1 mm = 10,000 positional units

Use the applicable working unit format to input precision distances and angles. Table 2-2 outlines examples of architectural units.

TABLE 2-2: ARCHITECTURAL UNITS: 1:12:1000 AND 1:12:8000

Distance	Key-in Formats		
2'-0"	2	2:0	:24
1'-6"	1.5	1:6	:30
3'-2 3/4"	3:2.75	3:2 3/4"	
8 1/2"	:8.5	:8 1/2	

Table 2-3 outlines examples of civil/metric units.

TABLE 2-3: CIVIL/METRIC UNITS: 1:10:100 AND 1:10:1000

Distance	Key-in Formats	
2'-0"	2	2:0
1'-6"	1.5	1:5
3.75'	3.75	3:7.5
6"	.5	:5

LOCKS

There are several lock tools available for controlling how your drawing points are processed. These locks can be accessed by selecting the Active Locks icon found on the status bar at the bottom of the application window. These settings are global to all models in the design file. These settings are saved with the file using the Save Settings command.

Grid Lock: Locks your data points to the defined grid points and restricts all data points and tentative points to the grid.

Level Lock: Locks you to the active level, restricting access to all other levels.

Boresite (3D only): Allows you to snap to elements at any depth in the design cube regardless of active Z-depth setting.

Annotation Scale: Forces text and dimension elements to scale automatically based on the model's drawing scale.

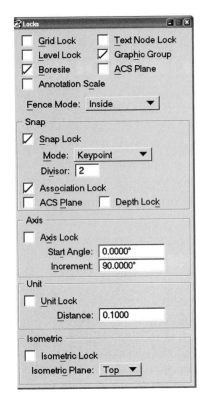

Text Node Lock: Restricts your text commands to text nodes only. This is rarely used today and is largely a legacy command.

Graphic Group Lock: Controls how elements grouped using the Add to Graphic Group command are managed. Pattern elements are automatically created as graphic groups.

ACS Plane (3D only): Locks your data points to the active Auxiliary Coordinate System Z-depth setting.

Fence Mode: Defines the active fence manipulation method: Inside, Overlap, Clip, Void, Void Overlap, or Void Clip. Refer to Chapter 6 for additional information on Fence commands and tools.

Snap Lock: Controls the functionality of the tentative snap. If inactive, the tentative capabilities are disabled. Refer to Chapter 5 for additional information on the Snap Mode and Divisor options.

Association Lock: Controls the "linking" capabilities of elements to each other. Elements can be linked to each other automatically using this lock setting.

ACS Plane (3D only): Locks your tentative points to the active Auxiliary Coordinate System Z-depth setting.

Depth Lock (3D only): Locks your data points to a specified Z value.

Axis Lock: Locks your cursor to a specified angle defined by the Start Angle and Increment Angle values. This restricts your ability to draw angles freely, similar to the Ortho command in AutoCAD.

Start Angle: Defines the start angle for calculating element axis restrictions.

Increment Angle: Defines the delta angles available from the start angle when using Axis Lock.

Unit Lock: Locks your data points to "invisible" grid points and restricts all data points and tentative points to this "invisible" grid. This is similar to grid lock except that the grid lock works with a "visible" grid.

Distance: Defines the distance between the "invisible" grid points.

Isometric Lock: Locks your data points to an isometric plane using top, left, and right; and restricts you to the angles of 30, 90, and 150.

Isometric Plane: Defines which plane the isometric lock is set to: top left or right.

TIP: *Use the Element Lock option to restrict modifications to graphics. You can access this command by using the key-in CHANGE LOCK.*

TEMPLATES AND SEED FILES

The concept of templates is widespread throughout many Windows applications and is probably not a new one to you. Even in AutoCAD you probably used a "template" file to define common settings one time and then inherit them from that point forward. A seed file in MicroStation is the same thing as a template. These seed files store common MicroStation settings that need to be carried through to all DGN files created from this point forward. Your organization should define corporate standards and define these standards in a custom seed file. There are two types of seed files provided: 2D and 3D. Many sample seed files are provided, and they vary mainly in the working unit setup.

A seed file is no different than any other DGN file you use to generate drawings, and is usually set aside with restricted access to maintain the integrity of the standards settings. The file extension for a seed file is *.DGN*, so theoretically any DGN file can be used as a seed file.

You can also use a DWG file as a seed file, which is generally recommended if your project requires DWG deliverables.

MODELS AND SHEETS

The V8 DGN file format is constructed of one or more models that can contain graphic elements. You can create two types of models in the V8 DGN file: a design model and a sheet model. A design model is the working model where most graphical elements will be placed and is available as 2D or 3D. A sheet model is used to compose assemblies and plotted output and is available in 2D or 3D.

This concept is similar to the ModelSpace/PaperSpace concept found in AutoCAD, but Bentley took it one step further to allow you to create multiple model spaces, and multiple paper spaces in a single DGN file.

MicroStation models can be compared to the sheets found in Excel in regard to their independence in both functionality and settings. In Excel, one sheet may show columns and rows formatted one way, while another sheet in the same workbook can use completely different columns, rows, and formatted data. Where Excel uses "textual" data, MicroStation uses "graphical" data such as lines, circles, and arcs.

MicroStation has independent graphical spaces called models. The number of models you can create is unlimited. *Just because you can create unlimited models does not mean, however, that you should.* So what are some of the uses for these new models? Use them to achieve the following.

❑ Improve your plotting environment

❑ Improve your working environment

❑ Organize your details

❑ Work with assembly drawings

❑ Control sheet size and annotation scale

Using Models

First, let's investigate how models are created and how you can navigate through them.

THE MODELS DIALOG

This button is available on the Primary Tools toolbar. However, it is hidden by default. You can activate the Models button by right-clicking directly on the toolbar and checking the option for Models. (Refer to Chapter 2 for additional interface information.)

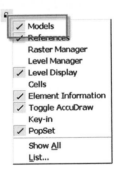

Create a Model. By default, the 2D design file will have one design model and no sheet model. You can use the Create Model button to create a new model of either type.

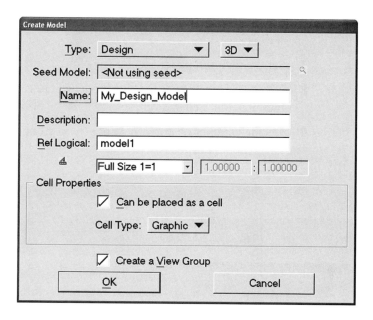

Design Models. The following are the steps for creating a design model.

1   Set the type of model to Design.

2   Select 2D or 3D.

3   Key in a model name and description (optional).

4   Key in a logical name if needed (optional).

5   Define the model scale if using annotation scale for text and dimensions.

6   Determine whether or not this model will be used as a cell in other DGN files. If being used as a cell, specify the type of cell as graphic or point.

7   Decide whether or not you want a view group created automatically for navigational purposes.

Sheet Models. The following are the steps for creating a sheet model.

1   Set the type of model to Design.

2   Select 2D or 3D.

3   Key in a model name and description (optional).

4   Key in a logical name if needed (optional).

5   Define the model scale if using annotation scale for text and dimensions.

6 If you prefer to "see" the paper edges, activate the Display Sheet Layout setting.

7 Set the required paper size for the plotted output.

8 Determine whether or not this model will be used as a cell in other DGN files. If being used as a cell, specify the type of cell as graphic or point.

9 Decide whether or not you want a view group created automatically for navigational purposes.

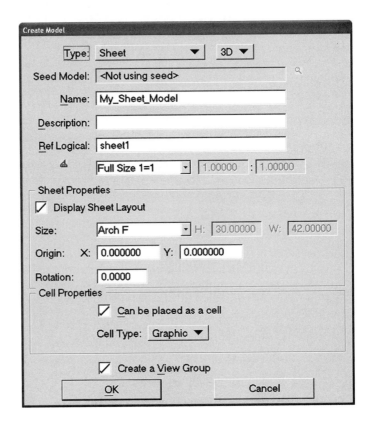

EDIT MODEL PROPERTIES

You can modify any of the model properties (with the exception of 2D or 3D) using the Edit Model Properties button.

IMPORT A MODEL

One of the best features available for models is the ability to import them from other design files. This is an excellent way of standardizing model configurations and easily distributing them throughout the organization.

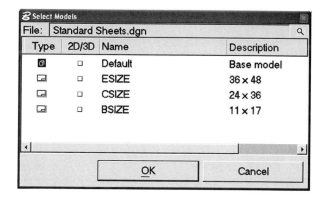

For example, your organization has standard paper sizes and borders. Set up a *Standard Sheets.DGN* file on your server and allow users to access it via the Import a Model button. Wouldn't it be a shame if you never had to set up another sheet? Table 2-4 outlines the content of a sample standard sheets file.

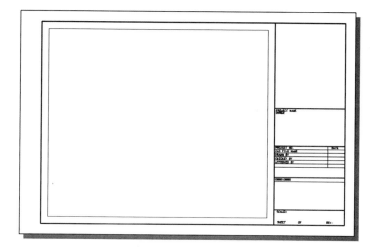

TABLE 2-4: STANDARD SHEETS FILE

DGN-specific Information	Model-specific Information
Levels	Working Units
Text Styles	Color Table
Dimension Styles	View Attributes

DGN-specific Information	Model-specific Information
Multi-Line Styles	Element Attributes
Active Angle	Grid Settings
Active Scale	View Settings
Coordinate Readout	Saved Views
Fence Mode	Named Groups
Snap Mode	
Locks	
Tag Sets	
Cell Library Attachments	
Line Style Resources	
AccuDraw Settings	

FILE TYPES

The following file types are supported in MicroStation V8.

DGN:	MicroStation design files
DWG:	AutoCAD drawing files
DXF:	Autodesk drawing exchange file
CEL:	Cell libraries
DGNLIB:	Standard content libraries for levels, text styles, dimension styles, and multi-lines
RDL:	Bentley redline files
S01:	Bentley sheet files
HLN:	Hidden line files
RSC:	Resource files and fonts
TBL:	Color tables and pen tables
CTB:	Autodesk color-based plot style tables

STB:	Autodesk named plot style tables
DEF:	File definitions
XLS:	MicroSoft Excel spreadsheets
CSV:	Microsoft Excel ASCII format
MDL:	Development languages application file
MA:	MDL compiled file
BAS:	Basic macro file
BA:	BAS compiled file
MVBA:	Visual Basic application file
MDB:	Microsoft database file
PAL:	Material palette file
MNU:	Menu file and function keys
PLT:	Plot driver files
SHX:	AutoCAD font files
INI:	Plotting settings files
UCF:	User configuration files
UPF:	User preference files
CFG:	Workspace configuration files
PCF:	Project configuration files
R01:	Interface resource files
M01:	Menu resource files
BCNV:	Batch convert settings files
BPRC:	Batch process settings files
PZIP:	Packager files
MAR:	Archive (V7) files
LIC:	License files

Raster-supported File Types

The following raster formats are supported in MicroStation V8.

```
Bil [ *.bil ]
CALS (Type I) [ *.cal;*.cals;*.ct1 ]
Compuserve GIF [ *.gif ]
ERMapper [ *.ecw ]
Georeferenced TIFF [ *.tif;*.tiff ]
HMR [ *.hmr ]
Img (24 Bit) [ *.a ]
Img [ *.p ]
Intergraph 29 [ *.c29 ]
Intergraph 30 [ *.c30 ]
Intergraph 31 [ *.c31 ]
Intergraph CIT [ *.cit ]
Intergraph COT [ *.cot ]
Intergraph RGB Compressed [ *.rgb ]
Intergraph RGB [ *.rgb ]
Intergraph RLE [ *.rle ]
Intergraph TG4 [ *.tg4 ]
Intergraph TIFF [ *.tif;*.tiff ]
Internet TIFF [ *.iTIFF ]
JPEG (JFIF) [ *.jpg;*.jpeg;*.jpe;*.jfif ]
JPEG 2000 [ *.jp2;*.j2k;*.j2c;*.jpc;*.jpx;*.jpf ]
MrSID [ *.sid ]
PCX [ *.pcx ]
PNG [ *.png ]
RLC [ *.rlc ]
Sun Raster [ *.rs;*.ras ]
TIFF [ *.tif;*.tiff ]
Targa [ *.tga ]
Windows BMP [ *.bmp;*.dib ]
▶ All Supported Raster Files
All Files (*.*)
```

3: View Control

CHAPTER OBJECTIVES:

❑ Learn to use multiple view windows effectively

❑ Learn to control the display of graphics

❑ Understand the purpose and use of view groups

❑ Learn to save typical view window settings

This chapter is intended to familiarize the AutoCAD user with the multi-view environment found in MicroStation. The understanding and use of multiple view windows are critical to improving your everyday productivity.

Learning to take advantage of more than one view window and how they can eliminate repetitive daily tasks is an important subject to master. MicroStation provides a multi-view environment that allows you to make use of as many as eight views simultaneously during the design process. Although these views work independently of one another, they allow you to work seamlessly between them. Think of them as eight independently controlled "cameras" that can each be used to view different areas of the same drawing.

The eight views also allow you to view a design file using different variations of levels, symbology, zoom area, view attributes, and so on. This flexibility can provide increased productivity when working with large geographic areas.

USING MULTIPLE VIEWS

AutoCAD users typically use a single viewport when working in model space. The availability of multiple viewports exists in AutoCAD, but most users utilize it only when working on a 3D model. The use of multiple views (see following figure) is a long-standing concept in MicroStation, whether you are working in 2D or 3D. One benefit found in MicroStation is the lack of the focus requirement when moving from one view window to another. When working in MicroStation, there is no concept of view focus and thus you are not required to move the focus by selecting a view. Remember, all views are available all of the time. Let's take a look at how you can put MicroStation's multiple-view functionality to work in a 2D drawing environment.

ACCESSING VIEW COMMANDS

Getting to the view control commands is simple because they are available from so many locations. This allows you to access them from a convenient location while working, and allows each user to use their preferred method.

View Control Toolbar

The View Control toolbar contains all of the 2D and 3D view control tools, as follows. Many of these buttons are 3D only and not available from a 2D design file.

- ❑ Update View
- ❑ Zoom Out
- ❑ Zoom In
- ❑ Window Area

- ❏ Fit View
- ❏ Pan View
- ❏ View Next
- ❏ Zoom In and Out (3D only)
- ❏ Set Display Depth (3D only)
- ❏ Show Display Depth (3D only)
- ❏ Camera Settings (3D only)
- ❏ Clip Volume

- ❏ Rotate View
- ❏ View Previous
- ❏ Copy View
- ❏ Change View Perspective (3D only)
- ❏ Set Active Depth (3D only)
- ❏ Show Active Depth (3D only)
- ❏ Render
- ❏ Clip Mask

View Control Buttons

The buttons shown in the following figure are available in the scroll bar area of each individual view. Some of these buttons are turned off by default, but each user can customize what buttons are available. Use the pulldown menu **Workspace > Customize** and the View Border tab to add additional view control buttons to the view border commands.

 UPDATE VIEW
The Update View tool will redraw or repaint a view window. This is usually needed when there has been significant editing done in a view and "pixel dust" is left behind after the edits are complete, or when edits are made in a view that has been dynamically panned.

 ZOOM IN
The Zoom In tool moves the view closer to the drawing, providing a more detailed view.

The Tool Settings dialog provides access to the zoom ratio setting, which allows you to zoom in faster or slower based on this value. The following are examples.

Zoom Ratio 2.00: Zooms in to display half the current view area, and elements appear two times larger.

Zoom Ratio 4.00: Zooms in to display one-fourth the current view area, and elements appear four times larger.

Zoom Ratio 1.00: Would make no change to the current view area, and element size would not change. This setting will effectively convert the Zoom tool to the Pan tool.

ZOOM OUT

The Zoom Out tool moves the view farther away from the drawing, providing a less detailed view. The Tool Settings dialog provides access to the zoom ratio, which allows you to zoom in faster or slower based on this value. The following are examples.

Zoom Ratio 2.00: Zooms out to display twice the current view area, and elements appear half the previous size.

Zoom Ratio 4.00: Zooms out to display four times the current view area, and elements appear one-quarter the previous size.

Zoom Ratio 1.00: Would make no change to the current view area, and elements size would not change. This setting will effectively convert the Zoom tool to the Pan tool.

WINDOW AREA

The Window Area tool allows you to specify the exact portion of the drawing you need to view. By specifying a rectangular area you can zoom in to an area of the drawing exactly as needed. The aspect ratio of the defined rectangular area is determined by the aspect ratio of the view from which the window area command was selected.

The Tool Settings dialog provides you with the ability to apply the resulting view to an alternate view window. If you select the Window Area tool from view 1, by default it is assumed you want to apply the result to view 1. If you select the Window Area tool from view 2, it is assumed you want to apply the result to view 2. By selecting a different "apply to window" in the tool settings dialog, you can make the area selection using one view and display that area result in another view.

TIP: *You can keep one window open, showing the entire design, and use it to perform viewing commands in other windows. This is similar to Aerial View in AutoCAD.*

Window Area only changes the region of the drawing being viewed, it does not change the display levels or view attributes in the destination view. If you select the Window Area tool from a toolbar or from a pop-up menu, MicroStation is unable to predict where you want to apply the result. In either case, you need to select the view to which you want to apply the result.

 FIT VIEW
The Fit View tool allows you to "fit" the entire drawing in a single view. The Tool Settings dialog provides options for optimizing what portion(s) of the drawing you want to fit in the view.

All: Fits all files, including active, reference, and raster attachments.

Active: Fits only the active design file.

Reference: Fits only reference file attachments.

Raster: Fits only raster file attachments.

 ROTATE VIEW
The Rotate View tool allows you to rotate the view of the drawing, but not the actual elements in the drawing, for easier viewing and manipulation.

The Tool Settings dialog provides two methods for rotating the view.

2 Points: Rotates the view interactively using a pivot point and angle.

Unrotated: Reorients the view back to the default "top" view.

AUTOCAD TIP: *You can use the RV shortcut key-in to rotate the view by a specific angle. Key in* RV=angle_value *and select the view you want to rotate. There are many more shortcut key-ins available in MicroStation (see Appendix A).*

PAN VIEW
The Pan View tool allows you to move around the drawing without modifying the zoom ratio.

The Tool Settings dialog provides an option for dynamic panning, which allows you to see the drawing move during the pan operation.

VIEW PREVIOUS
The View Previous tool allows you to back up to the previous view area, simulating an Undo of the last view manipulation. MicroStation stores the current session view manipulations in memory for possible undo later. View Previous will recognize changes made to view attribute settings including changes made to levels.

VIEW NEXT
The View Next tool allows you to move forward to the next view area, simulating a Redo of the last view manipulation. MicroStation stores the current session view manipulations in memory for possible redo later. The View Next command can be issued only after a View Previous has been used. This tool recognizes changes made to view attribute settings including changes made to levels.

COPY VIEW
The Copy View tool allows you to copy an existing view's settings to another view. All settings are copied, including viewing area, view attributes, and level settings.

Right-click Menu View Controls

The view control commands can also be accessed through a right-click menu. This method is recommended if screen real estate is limited, or if you rarely use a particular view command and only need to access it occasionally.

- ❑ Quickset Save
- ❑ Quickset Recall
- ❑ Displayset

Reset	
Quickset Save	▶
Quickset Recall	▶
Displayset	▶
Update View	
Fit Active Design	
Window Area	
Window Center	
Zoom In	
Zoom Out	
Previous	

There are some unique view control commands available from this menu that are not found on the View Command toolbar or View Control buttons.

AutoCAD Command Comparison

AutoCAD View Commands	MicroStation View Commands
Regen All	Update View
Zoom Window	Window Area
Zoom Extents	Fit View
UCS about Z with UCSFOLLOW ON	Rotate View
Zoom Previous	View Previous
Zoom Next	View Next
N/A	Copy View
3D Orbit	Change View Perspective (3D only)
3D Orbit—Adjust Clipping Plane	Set Display Depth (3D only)
N/A	Show Display Depth (3D only)
Camera	Camera Settings (3D only)
3D Orbit—Front/Back Clipping On/Off	Clip Volume

AutoCAD

RIGHT-CLICK MENUS

AutoCAD provides quick access to the view commands using a Right-Click menu; however, most users prefer access using mouse controls.

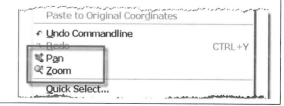

Mouse-activated View Controls

The most commonly used view commands are available right on the mouse. These settings can be changed via **Workspace** > **Preferences** > and select the *Mouse* category.

- ❑ Zoom In
- ❑ Zoom Out
- ❑ Pan

AutoCAD Command Comparison

AutoCAD
MOUSE-ACTIVATED VIEW COMMANDS Among the most productive tools in AutoCAD are the mouse tools themselves. Full access to these tools require that you have a two-button mouse with a wheel. However, some tools are available without the wheel.

AutoCAD
The mouse buttons and wheel in AutoCAD provide easy access to the following commands.
 	
    ❑ *Zoom In:* Roll wheel forward
    ❑ *Zoom Out:* Roll wheel backward
    ❑ *Pan:* Hold down the wheel
    ❑ *Dynamic Pan:* Hold down the wheel + hold down the Ctrl key on the keyboard.
    ❑ *Zoom Extents:* Double-click the wheel
MicroStation V8 also provides some of these tools on the mouse, with subtle differences.

AutoCAD		MicroStation
Mouse pick: Zoom In	Keyboard: *Zoom In*	This functionality is identical to that found in AutoCAD. Roll the wheel forward to zoom in. The location of the cursor prior to rolling the wheel controls the point that is zoomed in on.
Mouse pick: Zoom Out	Keyboard: *Zoom Out*	This functionality is identical to that found in AutoCAD. Roll the wheel backward to zoom out. The location of the cursor prior to rolling the wheel controls the point that is zoomed out from.

Mouse pick:　　Keyboard: *Pan* Pan	You cannot use the wheel to pan in MicroStation V8. However, there are several alternative options to choose from. First, try using the Pan View command located in the view control buttons found at the bottom of each view. Be sure to turn on the Dynamic Display option in the Tool Settings dialog to more closely mimic the pan found in AutoCAD. Second, you can use the typical MicroStation Pan command using a combination of the Shift key on the keyboard and Left mouse button. You must continue to hold down the Shift key and the Left mouse button while dragging the cursor around the screen. This Pan command will pan in the opposite direction from what you are used to in AutoCAD.
	DYNAMIC PAN There are user preference options that are somewhat similar to this function in AutoCAD, however, they are difficult to control at best. As a long time MicroStation and AutoCAD user I recommend the options previously discussed. To access these settings: **1**　Select **Workspace >Preferences**. **2**　Select the *Mouse* category and modify the Pan Radial and Pan with Zoom settings. Try it out—you might find you prefer this method (that is why they are "user preferences").
Keyboard: *ZOOMFACTOR* This is the system variable that controls the speed of mouse zooming capabilities in AutoCAD.	You can control this setting in MicroStation through the Zoom Ratio setting found in the Tool Settings dialog. You can permanently change the Zoom Ratio setting through user preferences.

AUTOCAD TIP: *Think of your cursor as a car driving along the screen. As you move the cursor on the screen you proceed forward in that direction, causing the drawing to pass you by. The farther you drag the cursor from the "start point," the harder you are pushing on the accelerator so the faster you can pan. If you keep this in mind, you "drive" the pan command to go where you want.*

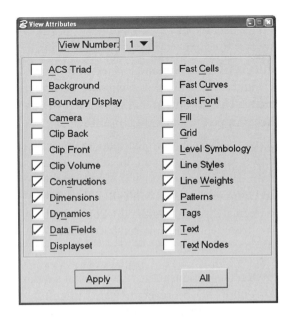

VIEW ATTRIBUTES

There are several attributes associated with MicroStation view windows that control everything from how text displays to fill patterns and drawing aids. These attributes (see the figure at left) control what type of design data, and at what level of detail, the user sees data in individual view windows. This information can be controlled independently for each view. For example, you can speed up the refresh rate of a view window by enabling fast text, fast cells, fast curves, and disabling patterns. The display of construction lines for design elements can also be controlled. The next section provides the definitions of all available view attributes.

The View Attributes dialog can be accessed in any of the following ways.

❑   Select **Settings > View Attributes**.

❑   Select the **Bentley "B" logo** on any view window and then select **View Attributes**.

❑   Press Ctrl + B to open the View Attributes dialog.

The following is a list of common view attributes and their definitions. Each user will develop their own preferences for controlling view attributes display, and these preferences can be saved with the design file. View attribute settings are not saved with the design file until the Save Settings command is executed.

ACS Triad: An XYZ coordinate triad similar to the UCS icon found in AutoCAD. This option displays at 0,0,0 in the current coordinate system.

Boundary Display: Displays the boundary lines for clipped references and models and helps to distinguish the extents of a reference file from other graphics. The type of boundary displayed is controlled by the Hilite Mode option, found in the Reference dialog.

Camera: Displays a camera angle of the view. Available in 3D only.

Clip Back: Controls the display of elements located behind the defined clipping area. Available in 3D only.

Clip Front: Controls the display of elements located in front of the defined clipping area. Available in 3D only.

Clip Volume: Controls the display of elements located outside a defined volume area. If activated, the view is restricted to elements within the defined volume area. If no clip volume has been defined, it has no effect.

Constructions: Displays construction class elements only. Any element can be placed as construction class rather than primary class. However, the default element class is primary.

Dynamics: Controls whether or not you see simple screen dynamics when elements are moved or modified.

Data Fields: Displays the underscore (_) character on enter-data field text elements so that they can be distinguished from normal text elements. Text displayed with characters and an underscore character are enter-data fields. The use of the underscore character to represent an enter-data field is controlled by the user preference setting found in **Workspace > Preferences > Text > ED Character.**

DisplaySet: Causes the view to display only the elements in a predefined selection set. All other element displays are disabled. This allows you to limit what graphics are displayed independent of levels and clipping boundaries.

Fast Cells: Displays all cells as "simple boxes" rather than in complete detail, and speeds up the display of the drawing.

Fast Curves: Displays complex curved elements as straight linear segments, and speeds up the display of the drawing.

Fast Font: Displays all text in the "fast font" rather than in specific font details, and speeds up the display of the drawing.

Fill: Controls the display of solid fills, commonly used for opaque shapes and filled fonts.

Grid: Controls the display of a drawing grid. The spacing of the grid is controlled by design file settings found in **Settings > Design File > Grid.**

Level Symbology: Controls the display of alternative element properties such as color, weight, and style. The alternative properties are defined using the Symbology Overrides settings in Level Manager.

Line Styles: Controls the display of custom line types in detail or as solid or continuous lines, and speeds up the display of the drawing.

Line Weights: Controls the display of element line weight. When deactivated, all line weights are displayed as weight 0.

Patterns: Controls the display of patterns or hatches that can slow down the display of the drawing.

Tags: Controls the display of tag elements, which are the similar to block attributes found in AutoCAD.

Text: Controls the display of text. When deactivated, no text will be displayed for both text elements and dimensions.

Text Nodes: Displays the node number and origin for multiline text elements. This node number is important to the internal workings of MicroStation.

ELEMENT CLASSIFICATIONS

The concept of element classifications dates to the processes and steps used in manual drafting. Manual drafting used to consist of linework and guidelines (construction lines), both used to create a final engineering drawing. The process of drawing construction lines was still used in the early days of CAD, when command functionality and precision input were more difficult to use. You probably won't use many construction lines today, but these legacy commands and functions remain in many CAD products.

All elements in MicroStation have a classification (class) rating assigned to them. The available element classes are as follows.

❑ Primary   ❑ Primary Rule

❑ Construction   ❑ Construction Rule

❑ Dimension   ❑ Linear Pattern

❑ Pattern Component

By default elements placed in a design file are considered primary class elements, which means that they are intended as permanent drawing components. Construction class elements can be placed in the design file when they are intended as temporary or are considered guidelines for creating additional primary elements.

Some unique capabilities are available when using MicroStation's construction class elements. You can control the display of construction lines without affecting primary lines, independent of level display. Construction lines are a view attribute that can be *activated* or *deactivated* in any view window.

You can also choose to plot primary class elements only. This allows you to place construction class elements in the design file for informational purposes only, and you can choose not to plot them. Use the print attribute settings available in plotting to accomplish this. You can change an element's classification using the Change Properties or Element Information tools (discussed in Chapter 6).

AutoCAD Command Comparison

AutoCAD
VIEW DISPLAYS AutoCAD does not provide equivalent view display controls when compared to MST. However, there are a few similarities. Available view display controls include the following: ❑ Grid ❑ Dynamic ❑ Lineweight SNAP GRID ORTHO POLAR OSNAP OTRACK DYN LWT MODEL

AutoCAD	MicroStation
Grid	Use the View Attributes Grid setting to display a grid in the appropriate view. Control the grid size using Design File settings. Access the grid settings using: the pulldown **Settings > Design File.** Select the *Grid* category and modify the grid settings as needed.
Dynamic This is a new setting in AutoCAD 2006 that provides on-screen feedback for most commands.	The only dynamic controls available in V8 are those that control whether you want to see graphics move with your cursor during modifications, such as copy, move, stretch and other modify commands. You can control this dynamic display using the View Attribute settings.
Lineweight The display of lineweight in AutoCAD has been available for many years, even though the typical AutoCAD user chooses not to use this feature.	You can disable the display of lineweight in MicroStation using View Attribute settings.

SAVE SETTINGS AND SAVED VIEWS

The Save Settings tool saves file- and view-specific settings, not the file data. These settings include view positions and display, design file settings, level settings, view attributes, and color table settings. The settings for all views will be saved. The Save Settings command is accessed by any of the following methods.

❑ Select **File > Save Settings.**

❑ Use the Ctrl + F shortcut key combination.

❑ Activate the automatic Save Settings feature found in MicroStation's user preferences. This causes the Save Settings command to be executed automatically whenever you close a file or exit the application. You can access this preference setting from **Workspace > Preferences > Operation > Save Settings on Exit.**

Saved views are a different concept and are very useful for automating day-to-day viewing manipulations, which can save considerable time down the road. Ask any user how much time they spend zooming and panning on a daily basis and you will soon see how this tool can eliminate repeated view manipulations.

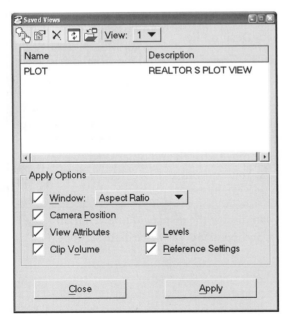

It is often worthwhile to spend time setting up a standard view with a window area, view attributes, and level state (on/off) so that you can then save it using the Saved Views command. This way, you do not have to repeat these steps again later. Everyone uses these view settings a little differently, and thus if another user opens your file and makes changes to the active view settings you can instantly recall your saved view settings. Saving a view is a simple process, as follow.

1　Set up the view area, levels, and symbology in the desired view window.

2　Select **Utilities > Saved Views**.

3　Key in a logical name for later recall.

You can also use saved views to automate reference file attachments and clipping boundaries while using the new model's functionality. Refer to Chapter 6 for additional information on reference files.

AutoCAD Command Comparison

AutoCAD
NAMED VIEWS I often wonder why AutoCAD users don't use named views more than they do. I consider this one of the most overlooked tools in the AutoCAD arsenal. But I am sure that when you see what additional functionality is available in MicroStation's Saved Views you won't be able to resist. So what are the differences between named views and saved views? The following is a list of what is saved with each view type. MicroStation's Saved Views store more information than their counterparts in AutoCAD. Saved Views can memorize window aspect ratios, size and position, camera position, view attributes, clip volume, levels, and reference settings.

AutoCAD Named Views	MicroStation Saved Views
Category: Defines what Sheet Set category	N/A
Location: Defines which model or layout	Window
Viewport: Defines any viewport association	N/A
Layers: Defines what layer state is stored	Levels
UCS: Defines what coordinate system to use	Window
Perspective: Defines if view is in perspective	Camera
N/A	View Attributes
N/A	Clip Volume
N/A	Reference Settings

Orthographic and Isometric Views

The following outlines respective MicroStation and AutoCAD predefined view orientation settings available by default.

AutoCAD Command Comparison

AutoCAD	
ORTHOGRAPHIC AND ISOMETRIC VIEWS This is a list of predefined view orientation settings that are available by default. Identical settings are available in MicroStation.	
AutoCAD Orientation Views	**MicroStation**
Top	Top
Bottom	Bottom
Front	Front
Back	Back
Left	Left
Right	Right
Southwest Isometric	Isometric
Southeast Isometric	Right Isometric
Northwest Isometric	Must be rotated manually
Northeast Isometric	Must be rotated manually
3D Orbit	Dynamic

VIEW GROUPS

Use the View Groups functionality to control view window arrangements and their associated display settings. These groups can store and recall the same information as saved views, with the addition of multiple view window control. View groups can also be used to navigate between models and sheets. In Exercise 3-1, following, you have the opportunity to practice using MicroStation view control functionality.

EXERCISE 3-1: CONTROLLING VIEWS

In this exercise you will learn how to control multiple views in MicroStation using basic view control commands such as View On, View Off, Zoom, Pan, Window Area, and Fit. Discover how you can take advantage of more than one view.

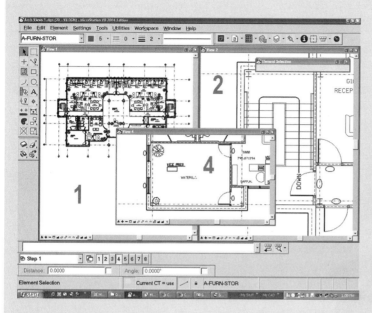

1 Open the design file *ARCH_VIEWS_1.DGN.*

The first thing you need to do is to turn views on or off as needed. As you can see upon opening the design file, there are three views open in this design configuration. You can turn them on or off using the View Groups toolbar located at the bottom of your application window.

2 Select the number 4 to turn off view 4. Select the number 3 to turn on view 3.

3 To rearrange the view windows, go to the pull-down menu **Window** > **Tile** to tile all three view windows.

The next few steps will teach you how to use the mouse view controls. First, let's use the wheel to zoom in and out.

4 Place your cursor in the middle of view 1 and roll the mouse wheel forward to zoom in and backward to zoom out. Note that the location of the cursor (crosshair) determines the point to zoom about.

AUTOCAD TIP: *If you issue a tentative snap in the view and then roll the wheel, the result is a combination of zoom center and zoom in/out functionality.*

Next, let's use the mouse to pan in a view.

5 Place your cursor in the middle of the view 2 window. Hold down the Shift key and the left mouse button.

Drag your cursor to the right and the view should pan. Move your cursor farther to speed it up

Practice moving around in the view with the dynamic pan functionality.

Now you are ready to use the View Control tools located on each view window.

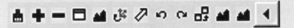

6   Select the Fit View tool from view 1. You should see immediately that the entire drawing is now in that view. This is similar to the Zoom Extents command in AutoCAD.

7   Issue another data point in view 2 and a Fit View command will execute there as well. (That is, you do not have to select the Fit View command from view 2.) Once you are in a view command, you can execute that view command in any open view window.

8   Select the Update View tool from view 3 and the view display will regenerate. It is almost instantaneous, so watch closely. You might see just a flicker.

We now want to learn to use the Window Area command to obtain the exact region needed in our view.

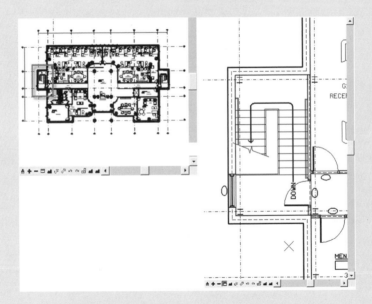

9   Select the Window Area tool from view 1.

10 Draw a box around the stair area on the west side of the building in view 1.

Now let's try to use the Window Area command between views.

11 Select the Window Area tool from view 3.

12 Draw a box around the stair area on the east side of the building in view 2.

The result of a window area command is displayed in the view window you selected the view command from. Thus, the result should end up in view 3.

Your final view windows should look like those shown in the following figure.

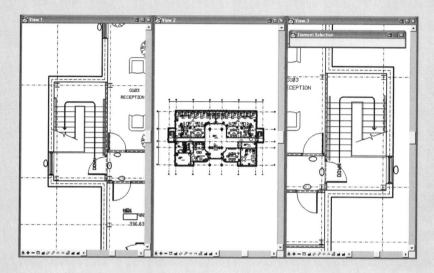

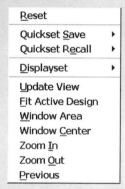

Believe it or not, there is one more option for accessing the view control commands. Because these tools are used so often every day, they are also available as a pop-up menu.

13   Place you cursor anywhere in a view window.

14   Hold down the Shift key and right-click.

You should see this pop-up menu on the screen containing the previously discussed view commands.

In Exercise 3-2, following, you have the opportunity to practice using Micro-Station's view attributes functionality.

EXERCISE 3-2: VIEW ATTRIBUTES

In this exercise you will learn how to establish what type of elements are used and how they display in view windows.

1   Open the design file *CIVIL_VIEWS_1.DGN*.

Let's find out how we can "tweak" the appearance of the view windows based on the view attributes.

2 You can access the View Attributes dialog via one of the following methods.

❑ Press Ctrl + B.

❑ Pick on the **Bentley "B"** on any view window and select the View Attributes command.

❑ **Settings > View Attributes**

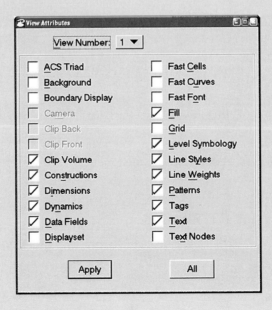

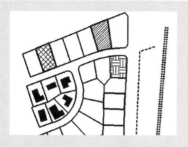

Project Number 2006-10243

Project Number 2006-10243

3 Toggle OFF the *Text* attribute and click on the Apply button to apply this change to view 1.

You should immediately see that all text has disappeared in view 1.

4 Toggle the *Text* attribute back ON.

5 Toggle the *Data Fields* attribute ON and OFF and apply it to view 1.

Note the changes in the Project Number display.

6 Toggle the Level Symbology attribute ON and OFF and apply it to view 1.

Note the changes in color, line styles, and line weights.

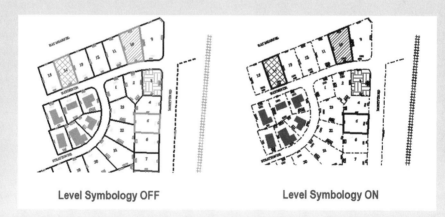

Level Symbology OFF   Level Symbology ON

7 Toggle the *Line Styles* attribute ON and OFF and apply it to view 1. Note the changes in the railroad line style located east of the parcels.

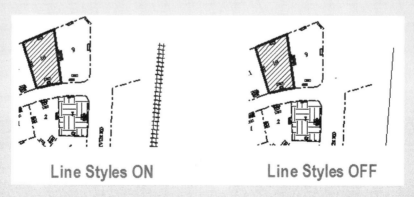

Line Styles ON   Line Styles OFF

As you can see, there are several beneficial display attributes that simplify your working view, your plotting view, or any viewing circumstances you might find yourself in.

AUTOCAD TIP: *Level symbology functionality in MicroStation is very similar to AutoCAD's VISRETAIN settings, but with a twist! These settings have an ON and OFF switch. Some of you might even compare this to a combination of VISRETAIN and layer states, and you are right.*

In Exercise 3-3, following, you have the opportunity to practice working with MicroStation's transparent view commands.

EXERCISE 3-3: TRANSPARENT VIEW COMMANDS

In this exercise you will learn how transparent the functionality between multiple views really is. You should be beginning to realize that even though there are eight independent views they really act like one.

While using the Line tool, we can run the zoom and window area commands transparently right in the middle of the drawing function. The dashed lines represent the lines you are about to complete.

1   Open the design file *TRANSPARENT_VIEWS.DGN*.

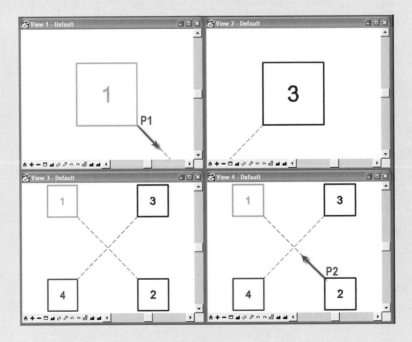

In the next few steps you will draw a line from BOX 1 to BOX 2 and use the Zoom In command in the middle of the line placement command.

2 Select the Line tool and snap to the keypoint (end point) at P1.

3 Select the Zoom In tool from the view controls in view 4. Issue a data point in the middle of BOX 2 to zoom in on the box.

4 Issue a reset (right-click) to complete the zoom command and return to the place line command.

5 Snap to the keypoint at P2. Issue a reset to complete the command.

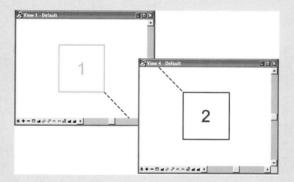

Let's try another one. Draw a line from BOX 3 to BOX 4 and use the window area command in the middle of the line placement command to zoom in on BOX 4.

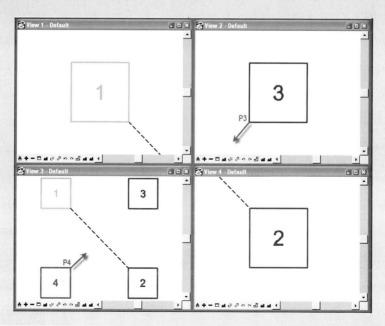

6   Select the Line tool and snap to the keypoint (end point) at P3.

7   Select the Window Area tool from the view controls in view 3. Draw a window around BOX 4 to move the display closer to the box.

8   Issue a reset (right-click) to complete the window area command and return to the place line command.

9   Snap to the keypoint at P4. Issue a reset to complete the command.

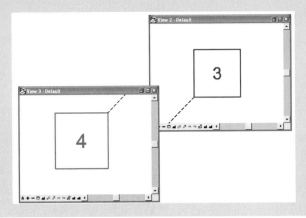

In Exercise 3-4, following, you have the opportunity to practice using saved views and view groups.

EXERCISE 3-4: USING SAVED VIEWS AND VIEW GROUPS

In this exercise you will learn how create a saved view with settings stored for later access. This is especially useful for taking advantage of level standards and plotting output standards.

1   Open the design file *SAVED_VIEWS_1.DGN*.

2   To access the Saved View command go to the pull-down menu **Utilities > Saved Views**.

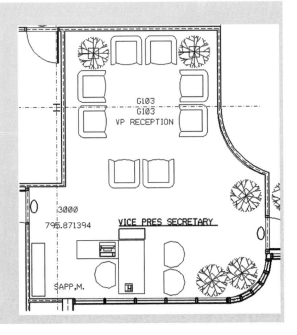

3 Click on the Save View button and key in the name *Reception*.

Click on the OK button to close this dialog.

4 Verify that the Levels setting is toggled on under the Apply options section of the dialog. (Your new saved view will remember the current status of the levels.)

Click on the Close button to exit the dialog.

5 To restore a previously saved view, go to the pull-down menu **Utilities > Saved Views**.

6 Highlight the saved view *Reception NO TEXT* and then click on the Apply button.

Note the changes in view 1. The text should have disappeared because those levels were not on when this saved view was created.

7 Highlight the saved view you just created named *Reception* and again click on the Apply button. The text should reappear because it was on when you saved the view.

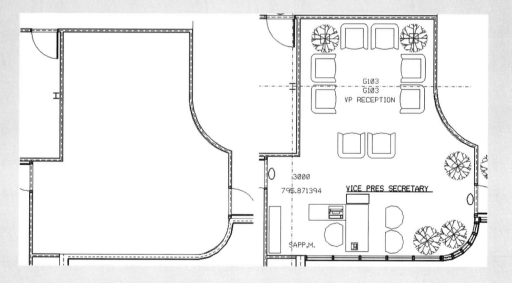

The last thing for you to learn is how to save and use view groups. This is new to MicroStation in V8. You will likely find some very beneficial uses for it.

8 Open the design file *VIEWGROUPS.DGN*.

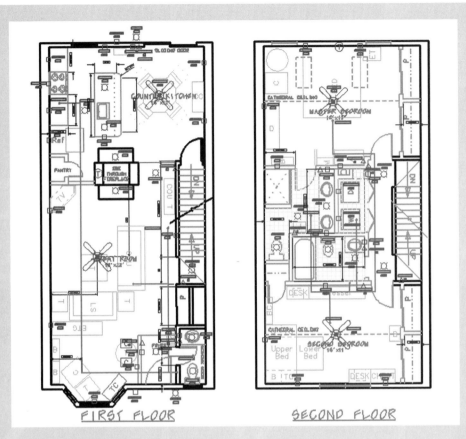

9 The View Groups toolbar is active by default, and you should see it docked at the bottom of the application window.

10 Select the View Group drop-down list and select Floor Plan from the list. Note the changes in the drawing levels.

11 Select the View Group drop-down list and select Electrical from the list.

This time you should see changes in the levels and changes in the view windows themselves. Views 1 through 4 are now open and tiled in the application.

12 Select the View Group drop-down list and select Default Views to return to the first view window setup.

You can make your own view groups by arranging the view in any order you want.

13 Turn off views 3 and 4. Tile the remaining view windows using **Window > Tile**.

14 Click on the Manage View Groups button.

15 Click on the Create View Group button and key in *My Views* for the name. Click on OK to close this dialog.

16 Switch to view group *Dimensions* and switch back to view group *My Views* to test the view group settings.

17 Click on the Close button to close this dialog.

USING 3D VIEWS

MicroStation separates 2D and 3D drafting using specific 2D and 3D design files. This minimizes the amount of information stored and reduces file size when 2D output is all that is required. This section discusses 3D viewing commands only.

Display Volume

Working in a 3D model can be difficult, especially when the model is large and complex. Getting to the "right spot" through all of the clutter can be confusing at best. Having the ability to clip the model to exactly what you want to work on is indispensable. Use the 3D clipping planes available via View Attributes to control the view of your working area.

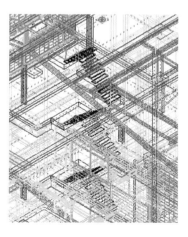
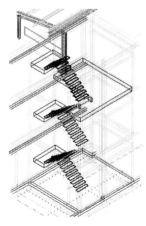

EXERCISE 3-5: 3D MULTIPLE VIEWS

The Clip Volume tool is used to isolate a specific 3D space within a 3D model for the purpose of viewing that space exclusively.

1   Open the design file *3Ddisc.DGN*.

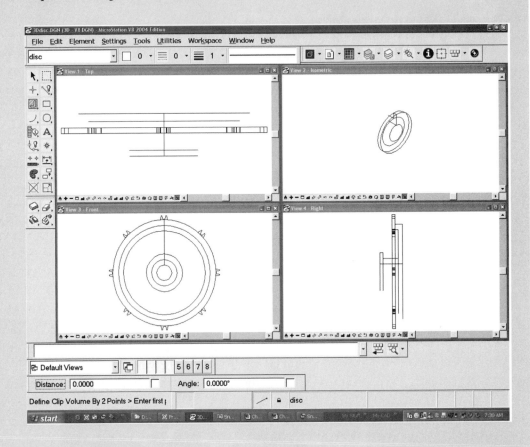

2   To access the Clip Volume toolbar, go to the pull-down menu **Tools > Tool Boxes.**

3   Select the 3D View Control toolbar.

4   Select the Clip Volume tool. The Tool Settings dialog will change to reveal a set of options for this command.

Select the Apply Clip Volume By 2 Points option identified as item 1, and uncheck the Display Clip Element option identified as item 2.

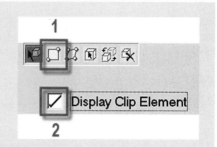

5 Pick around the hub of the disc at P1 and P2 in view 4.

Issue a data point anywhere in view 2 to apply the clip volume to view 2.

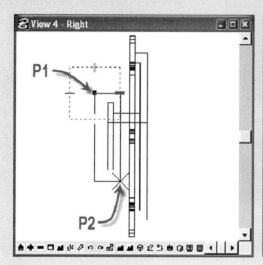

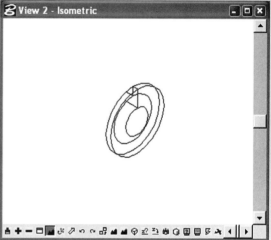

6 Use the Window Area tool to zoom in on the isolated hub.

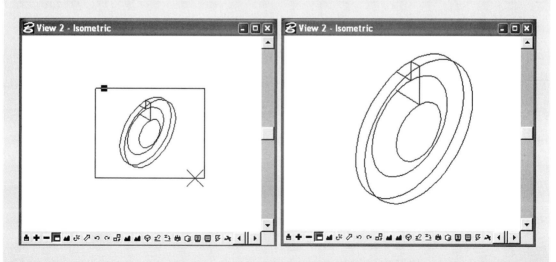

In the next few steps you will set up view 2 to contain a rendered view of the hub.

7 To access the Render View options, go to the pull-down menu **Utilities > Render > Smooth**.

Issue a data point anywhere in view 2 to apply the rendering attribute to the view. This is a temporary "rendering" of view 2 that can be removed using the Update View tool.

You can also use the Undo tool or the View Previous tool to return view 2 to its previous displayed state.

Try creating clipped volumes in other parts in this model. Select other destination views, and then render those views. This is another good example of how to take advantage of the flexibility in using MicroStation's multiple-view environment.

8 Close the file *3Ddisc.DGN*.

4: Basic Element Creation

CHAPTER OBJECTIVES:

❑ Learn the basics of AccuDraw

❑ Learn how to input precise graphics

❑ Learn to use snap tools efficiently

❑ Learn to use basic drawing tools to create lines, circles, arcs, and hatch patterns

This chapter introduces the basic drawing tools for creating lines, circles, arcs, and so on. In this chapter we focus on how to use the Tool Settings dialog to produce drawings with precision input. Learn to take advantage of your AutoCAD skills by applying them to MicroStation tools efficiently.

ACCUDRAW

The AccuDraw utility, introduced early in MicroStation J, is one of the best productivity enhancements added to MicroStation. This powerful tool has been enhanced several times since then and is better than ever in V8 2004. AccuDraw has matured into an indispensable tool for working efficiently in MicroStation. Users who are learning to use the AccuDraw tool for the first time should keep the following in mind.

Use it, but don't touch it!

It might appear that I am contradicting myself with that last statement. However, what I am trying to say is that AccuDraw's job is to predict what you are trying to do and perform accordingly. The secret to working with AccuDraw is to let it do its job, and to try not to get in the way. If you dock the AccuDraw dialog and ignore it, it really does work better. You might want to dock it by the status bar at the bottom of the application window, in that most of us tend to ignore command prompts anyway. The amazing thing about AccuDraw is how often it gets it right!

So, what is AccuDraw's job? AccuDraw tracks your cursor movement and tries to predict what you want to do. To take full advantage of this, look at it, read it, type in it, but don't touch it with the cursor. Let AccuDraw control the focus of what is going on in MicroStation, because this is a crucial aspect of how it works. If you touch the AccuDraw dialog using the cursor, you can move the focus to the wrong location and interfere with its predictability.

So, the secret to making AccuDraw easier to use is to *leave it alone*. Yes, you have to look at it and read it, but *don't touch it*. Touching is for later, when you have mastered the basics of AccuDraw. Once you try this utility you won't know how you ever lived without it, and AccuDraw in 3D is indispensable. If you have ever struggled with 3D planes, this utility is the answer to your prayers.

AutoCAD Command Comparison

The newer releases of AutoCAD provide input options such as Direct Distance Entry and Heads-Up Design. AutoCAD users that have adopted these new interface options should feel very comfortable with AccuDraw once the initial learning curve has been overcome. AutoCAD users still using the "old" @ key-ins rather than these new interface options may prefer to use the "old" MicroStation input options as well. They are discussed in the "Precision Input" section later in this chapter.

The Basics of AccuDraw

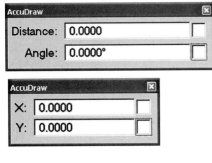

The following dialog must remain open in order to use AccuDraw. Place it in a convenient location, or dock it by the status bar in MicroStation.

(**TOP LEFT**): Polar AccuDraw (undocked)

(**BOTTOM LEFT**): Rectangular AccuDraw (undocked)

Rectangular AccuDraw (docked):

The key to drawing in MicroStation with precision is to learn to use the techniques provided by AccuDraw. Using AccuDraw is not required, but it simplifies precision input in a manner similar to that of AutoCAD's Polar, Otrack, and Direct Distance entry features.

The Compass

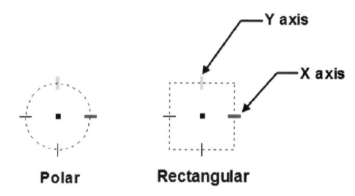

Polar   **Rectangular**

When moving in the positive or negative direction, AccuDraw tracks your movements and automatically assigns a positive or negative value to the key-in you provide. However, you need to pay attention to what you do with the cursor between commands because you might otherwise head off in the wrong direction with your key-in values. Don't worry; it just takes a little getting used to. The following AccuDraw shortcuts allow you to modify how the compass works.

1   Use the space bar to toggle between Rectangular and Polar modes.

2   Use the Polar (round) compass to enter distance and angle values.

3   Use the Rectangular (square) compass to enter X, Y, and Z coordinates.

Repeat Distance Indicator

When drawing with AccuDraw, it provides a "distance indicator" to help with the placement of repeated distances. The indicator is the perpendicular bar on the AccuDraw axis line.

Smart Lock

Another method to help automate your precision drafting is to "lock" to a specific X or Y axis by pressing the Enter key while on the preferred axis.

This prevents any movement off that axis and allows for quick "alignment" to other elements in the drawing.

AutoCAD Command Comparison

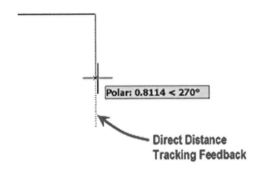

DIRECT DISTANCE ENTRY

The introduction of Direct Distance entry minimized the keyboard strokes required for precision input in AutoCAD. The dynamic screen feedback allows the user to easily "sketch" with precision and is very similar to the AccuDraw feedback provided in MicroStation.

HEADS-UP DESIGN

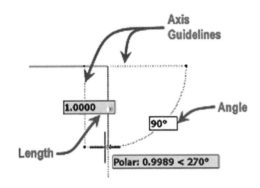

The introduction of the Heads-Up design feature in AutoCAD 2006 provides another input alternative and displays feedback previously displayed in the command line right on the cursor. Similar to AccuDraw's functionality, this live feedback provides input data for coordinates, lengths, and angles.

Reference Origin

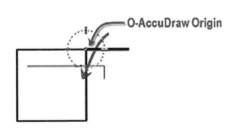

Use the letter O (Set Origin) keyboard shortcut to locate the AccuDraw compass at a "reference point" before keying in the distance or angle data. This shortcut specifically helps in the elimination of construction line drafting. Remember, if you are still drawing construction lines you should reevaluate your drafting process and consider how AccuDraw can eliminate those additional steps.

Rotate Quick

Use the RQ (Rotate Quick) keyboard shortcut to align with previously drawn graphics.

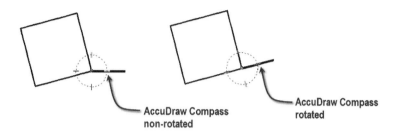

AccuDraw Compass non-rotated

AccuDraw Compass rotated

View Alignment

Use the V (View Rotation) keyboard shortcut to align the compass with the view, and the B (Base Rotation) keyboard shortcut to align the compass with the base axis.

AutoCAD Command Comparison

OTRACK

The introduction of Otrack and Polar tracking in AutoCAD provides a method of issuing graphical "reference points" for precision input. These temporary reference points allow the user to input coordinates or distances from existing objects, similar to AccuDraw's Reference Origin functionality. Temporary reference points are displayed on the screen as "blips" (shown in the figure following).

The AutoCAD user can then use this "Otrack blip" as a reference point for precision input. The following figure shows how you can reference an object endpoint and key in data referenced to that location.

OTRACK blip

Endpoint: 0.1541 < 0°

ACCUDRAW SHORTCUTS
Table 4-1 outlines shortcut keyboard entries for AccuDraw commands. Use these shortcuts for quick and easy AccuDraw input.

TABLE 4-1: SHORTCUT KEYBOARD ENTRIES FOR ACCUDRAW COMMANDS

Shortcuts	Name	Key-in	
Enter	SmartLock	AccuDraw Lock Smart	
Space	Change Mode	AccuDraw Mode	
O	Set Origin	AccuDraw SetOrigin	
V	View Rotation	AccuDraw Rotate View	
T	Top Rotation	AccuDraw Rotate Top	
F	Front Rotation	AccuDraw Rotate Front	3D
S	Side Rotation	AccuDraw Rotate Side	3D
B	Base Rotation	AccuDraw Rotate Base Toggle	
E	Cycle Rotation	AccuDraw Rotate Cycle	3D
X	Lock X	AccuDraw Lock X	
Y	Lock Y	AccuDraw Lock Y	
Z	Lock Z	AccuDraw Lock Z	3D
D	Lock Distance	AccuDraw Lock Distance	
A	Lock Angle	AccuDraw Lock Angle	
L	Lock Index	AccuDraw Lock Index	
RQ	Rotate Quick	AccuDraw Rotate Quick	
RA	Rotate ACS	AccuDraw Rotate ACS	
RX	Rotate about X	AccuDraw Rotate X	3D
RY	Rotate about Y	AccuDraw Rotate Y	3D
RZ	Rotate about Z	AccuDraw Rotate Z	
?	Show Shortcuts	AccuDraw Dialog Shortcuts	
~	Bump Tool Setting	AccuDraw Bump Tool Setting	
GT	Go to Tool Settings	Dialog ToolSetting	
GK	Go to Keyin	Dialog CmdBrowse	

Shortcuts	Name	Key-in	
GS	Go to Settings	AccuDraw Dialog Settings	
GA	Get ACS	AccuDraw Dialog GetACS	
WA	Write to ACS	AccuDraw Dialog SaveACS	
P	Point Keyin (single)	Point Keyin Single	
M	Point Keyin (multi)	Point Keyin Multiple	
I	Intersect Snap	Snap Intersect	
N	Nearest Snap	Snap Nearest	
C	Center Snap	Snap Center	
K	Snap Divisor	AccuDraw Dialog SnapDivisor	
U	AccuSnap Suspend	AccuSnap Suspend	
J	Toggle AccuSnap	AccuSnap Toggle	
Q	Quit AccuDraw	Accudraw Quit	

In Exercise 4-1, following, you have the opportunity to practice using AccuDraw's basic functionality.

EXERCISE 4-1: THE BASICS OF ACCUDRAW

In this exercise you will learn to use AccuDraw to draw a simple series of shapes.

1 Open the design file *ACCUDRAW_1.DGN*.

This exercise requires that AccuDraw be open, and it is recommended that you dock it at the bottom of the application window. Remember:

Use it, but don't touch it!

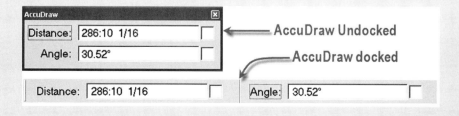

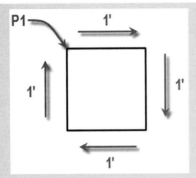

First, let's draw a simple 1' x 1' square shape using the Line command.

2 Select the Line tool and issue a data point anywhere in view 1. This point is labeled P1 in the figure at left. Verify that AccuDraw is using the Polar compass (round). If not, press the space bar to change it.

3 Drag the cursor in the X direction (→) and key in the value of *1*. The focus of AccuDraw was automatically in the Distance field, so your key-in was interpreted as a distance of 1.

Keep the cursor lined up with the X axis and issue a data point to accept this location.

4 Drag the cursor in the Y direction (↓) and key in the value of *1*. Again, the focus of AccuDraw was automatically in the Distance field, so your key-in was interpreted as a distance of 1.

Keep the cursor lined up with the Y axis and issue a data point to accept.

5 Drag the cursor in the X direction (←) and instead of using a key-in value of 1 let's use the Repeat Distance indicator to retrieve the previous value of 1.

When you see the Repeat Distance indicator, keep the cursor lined up with the X axis and issue a data point to accept.

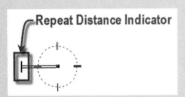

Repeat Distance Indicator

HINT: The Repeat Distance indicator looks like the figure at left.

6 Drag the cursor in the Y direction (↑) until you see the Repeat Distance indicator, and then issue a data point to complete the square.

Next, using the SmartLine command we will draw the same shape, except at a 45-degree angle.

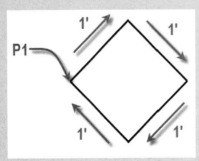

7 Select the SmartLine tool and issue a data point anywhere in view 1. This point is labeled P1 in the figure at left.

8 Drag the cursor at an angle of 45 degrees (↗) and key in the value of *1*.

We want to control the angle, so use the Tab key to move the AccuDraw focus to the Angle field. Key in the value of *45*. Issue a data point to accept.

9 Drag the cursor at an angle of 315 (-45) degrees (↘) along the rotated axis until you see the Repeat Distance indicator, and then issue a data point to accept.

10 Drag the cursor at an angle of 225 degrees (↙) until you see the Repeat Distance indicator, and then issue the last data point to complete the shape.

11 Drag the cursor at an angle of 135 degrees (↖) until you see the Repeat Distance indicator, and then issue a data point to accept.

In Exercise 4-2, following, you have the opportunity to practice using AccuDraw shortcuts.

EXERCISE 4-2: USING ACCUDRAW SHORTCUTS

In this exercise you will learn to use several of basic AccuDraw shortcuts. Learn these shortcuts and they will allow you to quickly perform the most-used functions of AccuDraw.

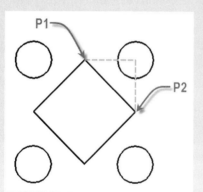

1 Open the design file *ACCUDRAW_2.DGN*.

We will add four circles to this shape using the AccuDraw Smart Lock and Set Origin shortcut key-ins. To place these circles without AccuDraw you would need to draw the dashed construction lines, but with AccuDraw we can place the circle where it belongs immediately.

2 Select the Circle tool and place a tentative point at P1. Verify that the Circle tool's method is set to Center.

HINT: A tentative point is issued with the middle mouse button or by pressing the right and left mouse buttons simultaneously.

3 Key in the letter *O* to activate the Set Origin shortcut. You should see the AccuDraw compass move to point P1.

4 Drag the cursor in the X direction (→) and press the Enter key to activate AccuDraw's Smart Lock command.

HINT: Smart Lock locks you to the AccuDraw axis.

5 Move the cursor to P2 but do not pick this point immediately. Instead, "hover" over P2 until the AccuDraw Alignment indicator displays. See the figure on the following page.

Only when you see this alignment indicator should you issue a data point to accept.

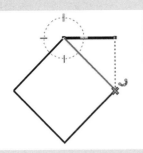

6 Key in a radius of *.25* to complete the circle command.

7 Complete the remaining three circles for this drawing.

HINT: Use the Repeat Distance indicator for the remaining circles.

Next, we will use the Smartline command to draw the capsule shape above using AccuDraw to simplify the drafting steps.

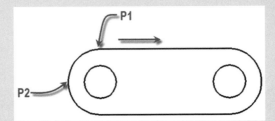

8 Select the SmartLine tool and issue a data point anywhere in view 1. This point is labeled P1 in the figure above.

Drag the cursor in the X direction (→) and key in the value of *2* for the length of the line segment. Issue a data point to accept.

9 Drag the cursor in the Y direction (↓) and press the tilde key (~) on the keyboard to toggle the Segment Type tool setting to Arcs. Key in the value *.5* for the radius distance of the arc segment. Issue a data point to accept.

If the arc is "swinging" in the wrong direction, use the cursor to change the swing angle. This can be accomplished by tracing the cursor along the current arc segment until it is "swinging" in the correct direction.

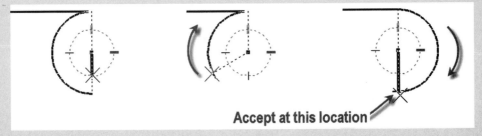

Line up the AccuDraw compass with the Y axis (↓) and issue a data point to accept.

10 Drag the cursor in the X direction (←) and press the tilde key (~) on the keyboard to toggle the Segment Type tool setting to Lines. Key in the value *2* for the length of the line segment.

11 Drag the cursor in the Y direction (↑) and press the tilde key (~) on the keyboard to toggle the Segment Type tool setting to Arcs. Key in the value *.5* for the radius distance of the arc segment. Issue a data point to accept and close the shape with another data point to complete the capsule.

Yes, you could snap to the center of the arc to place the circles, but our focus in this exercise is to learn key concepts of AccuDraw. Using this "simple" shape problem you can easily practice these important shortcuts.

12 Select the Circle tool and place a tentative point at P1. Verify that the Circle tool's method is set to Center.

13 Key in *O* (Set Origin) and the AccuDraw compass will move to P1.

14 Drag the cursor in the Y direction (↓) and press the Enter key to activate the AccuDraw SmartLock.

15 Move the cursor to P2 but do not pick this point immediately. Instead, "hover" over P2 until the AccuDraw Alignment indicator displays.

Only when you see this alignment indicator should you issue a data point to accept.

16 Key in a radius of *.25* to complete the circle command.

17 Complete the remaining circle for this drawing.

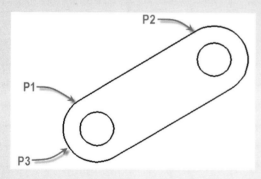

Next, we will use the Circle tool to add two circles to a rotated capsule shape, and use AccuDraw's Rotate Quick, View Rotation, and Set Origin options as shortcuts.

18 Select the Circle tool and place a tentative point at P1. Verify that the Circle tool's method is set to Center.

19 Key in *O* and the AccuDraw compass will move to point P1. However, there is a problem with the compass orientation to the capsule.

20 Key in the letters *RQ* to activate the Rotate Quick shortcut and pick the keypoint at P2.

21 Drag the cursor in the Y direction (↘) and press the Enter key to activate the AccuDraw SmartLock.

22 Move the cursor to P3 but do not pick this point immediately. Instead, "hover" over P3 until the AccuDraw Alignment indicator displays.

Only when you see this "alignment" line should you issue a data point to accept.

23 Key in a radius of .25 to complete the circle command.

24 Complete the remaining circle for this drawing.

In Exercise 4-3, following, you have the opportunity to practice using AccuDraw's mathematical calculations.

EXERCISE 4-3: USING ACCUDRAW CALCULATIONS

Finally, we will use AccuDraw to derive mathematically calculated locations. In this example we want to draw a center line between two existing lines.

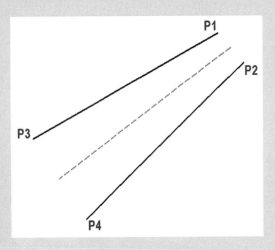

1 Select the Line tool and place a tentative point at P1.

2 Key in *O* and the AccuDraw compass will move to point P1.

3 Move the cursor to P2, but do not pick this point immediately.

Examine (look, but don't touch) the AccuDraw dialog and note that the focus is displaying the distance between P1 and P2.

Key in the characters */2* to divide the distance by 2, and then issue a data point to accept. Did you see the hidden calculator appear?

4 Place a tentative point at P3.

5 Key in *O* and the AccuDraw compass will move to point P3.

6 Move the cursor to P4 and key in */2* to divide the distance between P3 and P4. Issue a data point to accept.

An easy way to draw a centerline without using any construction lines, right?

PRECISION INPUT

The following provide examples of using coordinate, relative, and polar input to create lines.

Coordinate Input

To draw a 2'-6" line at 0 degrees, starting at 0,0 use the following key-ins.

XY = 0,0	and	XY = 2.5 (representing 2.5')
	or	XY = 2:6 (representing 2'-6")
	or	XY =: 30 (representing 30")

Relative Input

To draw a 2'-6" line at 0 degrees, starting at 0,0 use the following key-ins.

XY = 0,0	and	DL = 2.5 (representing 2.5')
	or	DL = 2:6 (representing 2'-6")
	or	DL = :30 (representing 30")

Polar Input

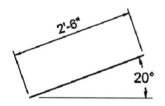

To draw a line at 20 degrees, starting at 0,0 use the following key-ins.

XY = 0,0	and	DI = 2.5,20 (representing 2.5' and 20 degrees)
	or	DI = 2:6,20 (representing 2'-6" and 20 degrees)
	or	DI = :30,20 (representing 30" and 20 degrees)

AUTOCAD TIP: *You can use the semicolon (;) in place of the colon (:) for distances in MicroStation. Why is this important? The reason is that it is easier to key in a semicolon in place of the colon because it does not require using the Shift key.*

AutoCAD Command Comparison

COORDINATE INPUT

To draw a 2'-6" line starting at 0,0 use the following key-ins.

0,0	and	2.5' (representing 2.5')
	or	2'6 (representing 2'-6")
	or	30 (representing 30")

RELATIVE INPUT

To draw a 2'-6" line at 20 degrees, starting at 0,0 use the following key-ins.

0,0	and	@2.5',0 (representing 2.5' in the X direction and 0 in the Y direction)
	or	@2'6,0 (representing 2'-6" in the X direction and 0 in the Y direction)
	or	@30,0 (representing 30" in the X direction and 0 in the Y direction)

POLAR INPUT

To draw a 2'-6" line at 20 degrees, starting at 0,0 use the following key-ins.

0,0	and	@2.5'<20 (representing 2.5 and 20 degrees)
	or	@2'6<20 (representing 2'-6" and 20 degrees)
	or	@30<20 (representing 30 and 20 degrees)

SNAP MODES

MicroStation and AutoCAD both offer specific methods of snapping to elements. Although these snap methods are more similar than different, there are some significant differences worth mentioning.

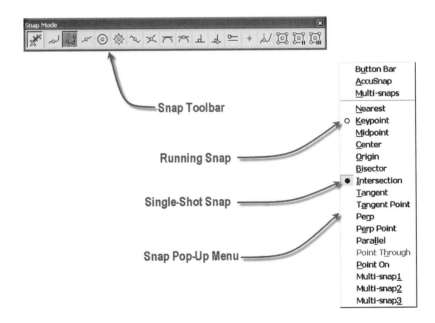

Snap Definitions

Table 4-2 outlines MicroStation snap mode definitions.

TABLE 4-2: MICROSTATION SNAP MODE DEFINITIONS

MicroStation Snap Mode	Description
Nearest	The closest point to the cursor on any element. There is no geometric significance to this point.
Keypoint	Predefined points on elements based on the Keypoint Divisor setting.
Midpoint	The midpoint of an element segment.
Center	The center (centroid) of an element.
Origin	The justification point of text elements, and the insertion point of cells.
Bisector	The midpoint of an entire element independent of segments.
Intersection	The point of intersection between two elements.
Tangent	The point tangent *to* an element.
Tangent Point	The point tangent *from* an element.
Perpendicular	The point perpendicular *to* an element.
Perpendicular Point	The point perpendicular *from* an element.
Parallel	Draw parallel to an element.
Point Through	A defined point to pass through.
Point On	Constrain an element to begin or end on an element.
Multi-Snap1, 2, 3	Saved combinations of the Keypoint, Center, Intersection, Origin, Bisector, Midpoint, and Nearest snap methods.

AutoCAD Command Comparison

OSNAPs

Table 4-3 outlines OSNAPs available in AutoCAD and their equivalent snap methods in MicroStation. However, there are a few that have no obvious match.

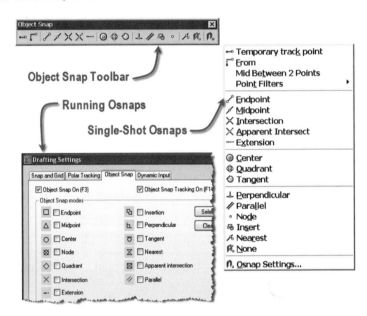

TABLE 4-3: AUTOCAD OSNAPS AND MICROSTATION EQUIVALENTS

AutoCAD OSNAP	MicroStation Equivalent
Endpoint	Keypoint
Midpoint	Keypoint or Midpoint
Center	Keypoint or Center
Node	Keypoint
Quadrant	Keypoint
Intersection	Intersection
Extension	Use AccuDraw's origin to simulate extension
Insertion	Origin
Perpendicular	Perpendicular
N/A	Perpendicular Point
Tangent	Tangent
N/A	Tangent Point

AutoCAD OSNAP	MicroStation Equivalent
Nearest	Nearest
Apparent Intersection	Intersection
Parallel	Parallel
N/A	Point Through
N/A	Point On

Snaps Accessibility

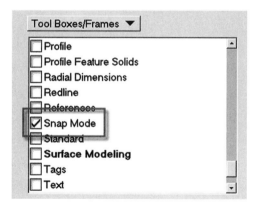

There are several ways to access the various snap methods via toolbar and pop-up menus.

To use the toolbar:

1 Select **Tools > Tool Boxes**.

2 Scroll to the Snap Mode toolbox and activate it to display the Snap toolbar.

Alternatively, perform the following.

1 Click on the Active Snap Mode icon located on the status bar at the bottom of the MicroStation application window.

2 Select Button Bar at the top of the pop-up.

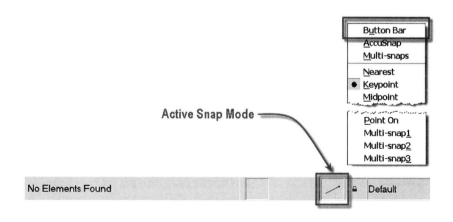

To access the pop-up menu:

1 Click on the Active Snap Mode icon located on the status bar at the bottom of the MicroStation application window.

2 Select the desired snap mode.

Alternatively, perform the following.

1 Hold down the Shift key and tentative click anywhere in the drawing window.

2 Select the desired snap mode.

One of the most unique and productive snaps found in MicroStation is the Keypoint snap mode. This snap method simulates many of the snaps available in AutoCAD, such as Endpoint, Midpoint, Center, and Quadrant. The keypoints on elements vary depending on the element type. These keypoints are controlled by the Keypoint Divisor setting, which is set to 2 by default. Using the default setting of this divisor function, each element segment is divided by 2. This creates the keypoints displayed in the following figure for each element type.

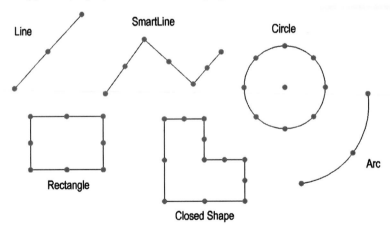

The keypoint divisor can be modified at any time to change the location of the keypoints on an element.

Keypoint Divisor = 1

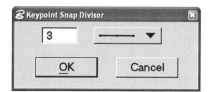

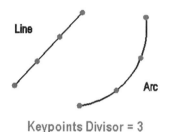

Line

Arc

Keypoints Divisor = 3

To change the Keypoint Divisor setting:

1 Using AccuDraw, key in *K* to access the Keypoint Divisor tool.

2 Key in *3* to modify the divisor setting.

Alternatively, perform the following.

1 Click on the Active Locks icon located on the status bar at the bottom of the MicroStation application window.

2 Select Full to access the Locks dialog.

3 Modify the Snap > Divisor setting.

Running Snaps

In MicroStation, a snap setting is always on and available from all commands, similar to the running snap setting in AutoCAD.

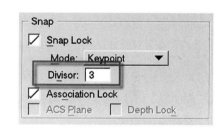

> **TIP 1:** *You can double click on the toolbar buttons to set a snap setting as a running snap. The associated button's appearance will change to display a "dotted" background to indicate the running setting.*

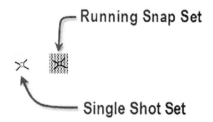

Running Snap Set

Single Shot Set

> **TIP 2:** *You can hold the Shift key down while selecting the snap from the pop-up menu to set it as a running snap.*

Single-Shot Snaps

A single-shot snap is only available for the next single snap operation. After this snap operation is executed, the snap setting will be reset back to the running snap. An example would be if you wanted to snap to keypoints on most of the elements but you needed an intersection snap for one snap only.

Multi-Snaps

Multi-snaps are a new addition to MicroStation. They allow you to set more than one running snap at a time. Now you can set your most commonly used snap modes to be running all the time, not just the keypoint snap. To use the Multi-snaps function you must configure what snaps you want to be available. MicroStation provides three multi-snap configurations. To define multi-snap configurations:

1 Click on the Active Snap Mode icon located on the status bar at the bottom of the MicroStation application window.

2 Select Multi-Snaps to access the Multi-Snap Set dialog.

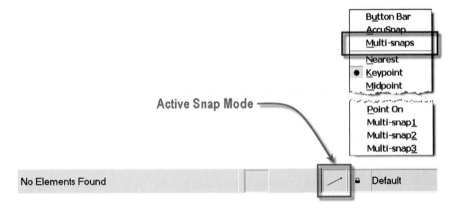

3 Click on one of the three configuration buttons and select which snap modes you want to have available. You can select as many as you need, but selecting all of them could be counterproductive. Too many snap modes can cause conflicts when trying to snap to specific points on elements.

Once you have defined the multi-snap configurations, you can use the multiple snap modes described in the following section to set them as running snaps.

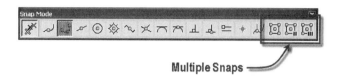

Multiple Snaps

Tentative Snap

Some of the snap modes in MicroStation require the use of tentative snap. Refer to Chapter 1 if you need more information on how to execute the tentative snap.

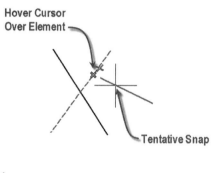

Hover Cursor
Over Element

Tentative Snap

APPARENT INTERSECTION SNAPS
When snapping to intersections that do not physically cross, you must issue a tentative snap on the first element and hover over the second element. MicroStation will display the calculated intersection of the two elements. Continue to hover the cursor over additional elements until the correct intersection is displayed. Issue a data point to accept the calculated intersection.

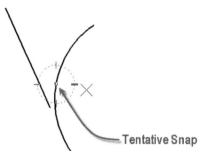

Tentative Snap

TANGENT SNAPS
The tangent snap also requires a tentative point to allow you to draw an element tangent to another element, or tangent from an element. Issue a tentative on the element you want to draw tangent to and from.

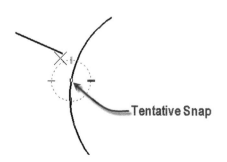

Tentative Snap

PERPENDICULAR SNAPS
The perpendicular snap requires a tentative point to allow you to draw an element perpendicular to another element, or perpendicular from an element. Issue a tentative on the element you want to draw perpendicular to or from.

In Exercise 4-4, following, you have the opportunity to practice using MicroStation's basic editing functions.

EXERCISE 4-4: PRECISION INPUT

In this exercise you will learn to input precise graphics using the Tool Settings dialog and the Key-in browser.

1   Open the design file *BASIC_INPUT.DGN*.

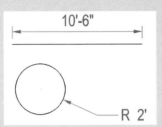

Using the Tool Settings Dialog

The next few steps use the Tool Settings dialog to draw elements using exact sizes and angles.

2   Select the Line tool and set the following settings in the Tool Settings dialog.

> *Length:*   10:6
> *Angle:*   0°

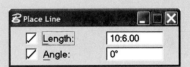

Note that the 10'-6" line at an angle of 0 is attached to the cursor.

3   Issue a data point in view 1 to place the line.

4   Select the Circle tool and set the following settings in the Tool Settings dialog.

> *Method:*   Center
> *Radius:*   2:0.00

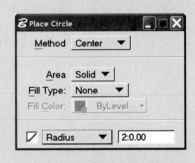

Note the 2'-0" radius circle attached to the cursor.

5   Issue a data point in view 1 to place the circle.

Using the Key-in Browser

The next few steps use the Key-in browser to draw the same elements using exact sizes and angles.

6   Select the Line tool and clear all settings in the Tool Settings dialog.

Key in *XY=0,0* in the Key-in browser to start the line at the coordinate 0,0.

7 Key in *DL=10:6,0* in the Key-in browser to draw the line 10'-6" in the X direction and 0 in the Y direction. Alternatively, key in *DI=10:6,0* in the Key-in browser to draw the line 10'-6" in length at the angle of 0.

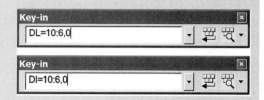

8 Select the Circle tool and clear all settings in the Tool settings dialog.

Key in *XY=1,1* to start the circle at the coordinate 1,1.

9 Key in *DL=2* in the Key-in browser to finish the circle with the radius of 2'.

In Exercise 4-5, following, you have the opportunity to practice using precision snaps.

EXERCISE 4-5: USING PRECISION SNAPS

In this exercise you will learn to use the basic snap methods available for precision drawing.

1 Open the design file *BASIC_SNAPS.DGN*.

The next few steps use the Snaps toolbar to snap to elements on the screen.

2 Select the Line tool and use the Center snap to snap to the center of the circle at P1.

Use the Intersection snap to snap to the intersection of the two lines at P2.

Reset to end the Line command.

3 Use the Keypoint snap to snap to the center of the circle at P1.

Use the Keypoint snap to snap to the vertex at P3.

Reset to end the Line command.

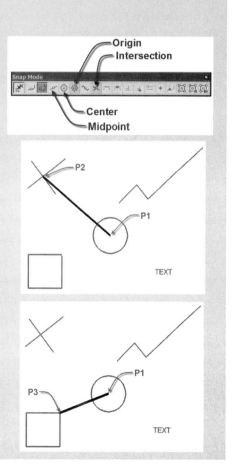

4 Use the Keypoint snap to snap to the center of the circle at P1.

Use the Origin snap to snap to the justification point of the text at P4. You can also use the Keypoint snap here.

Reset to end the Line command.

5 Use the Keypoint snap to snap to the center of the circle at P1.

Use the Midpoint snap to snap to the midpoint of the SmartLine segment at P5. You can also use the Keypoint snap here.

Reset to end the Line command.

The power of the Keypoint snap should now be obvious.

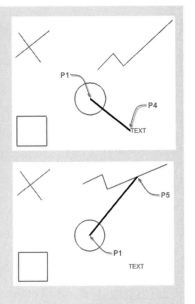

DRAWING COMMAND BASICS

Now that you can use AccuDraw for basic precision input, you can use it with all of your drawing and modification commands.

Tool Settings

Use the Tool Settings dialog (discussed in Chapter 1) to control how individual commands work and to take advantage of the options available.

DRAW A SMARTLINE

Using the SmartLine tool, instead of the Line tool, will reduce the drawing steps required, but more importantly it will reduce the steps needed later for modifications and manipulations. SmartLine elements are similar to polyline objects found in AutoCAD in that they can contain both straight and curved segments, but SmartLines can also contain "smart" vertices with automatic fillets, chamfers, or sharp corners. "Smart" vertices allow you to draw complex shapes more quickly and to modify them more easily.

You can modify the radius for fillets, chamfers, and sharp corners with "smart" vertex tool settings without having to clean up or delete old graphic pieces. You can also modify SmartLine segments without the need for tools such as Stretch, Lengthen, Otrack, Ortho, or Polar.

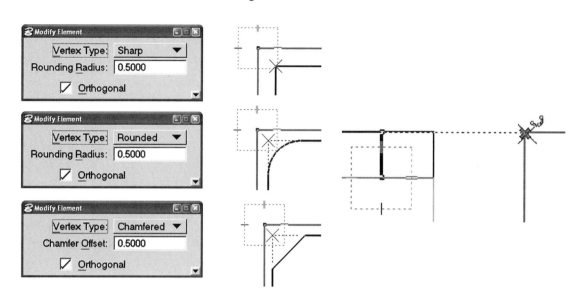

AutoCAD Command Comparison

AutoCAD	MicroStation
Mouse pick: Keyboard: *PLine* Polyline	Use the SmartLine tool to draw "polyline-like" objects in Micro-Station.
Line Options	Use the Smart Line tool and the Line segment option.
Halfwidth	Use the *element weight* attribute to get the "thick" line appearance.
Length	Use AccuDraw to define specific line and arc lengths and sizes.
Undo	Use the Undo and Redo tools to undo while running the command. By default, they are transparent commands in MicroStation.
Width	Use the *element weight* attribute to get the "thick" line appearance. For a varied line width in polylines, use a custom line style in MicroStation.
Close	Snap to the "first" point to close the SmartLine shape.

AutoCAD	MicroStation
Arc Options	Use the SmartLine tool and the Arc segment type.
Angle	Use AccuDraw in Polar mode to define the arc angle.
Center	Use AccuDraw to define the center point of the arc segment.
Close`	Snap to the "first" point to close the SmartLine shape.
Direction	Using AccuDraw, you can draw in any direction.
Line	Use the Lines segment type. Use the tilde key (~) on the keyboard to easily toggle between line and arc segment types.
Radius	Use AccuDraw to define the arc radius.
Second Pt	This option is not available from within the SmartLine tool.

DRAW A LINE

Use the Line tool for simple single line elements that do not need to be connected. The Tool Settings dialog provides the basic options such as Length and Angle. Using AccuDraw in conjunction with these tool settings will simplify the line creation process.

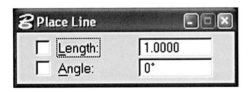

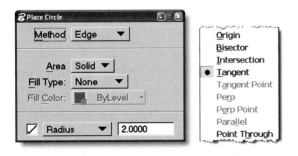

Undo (Undo a Specific Point). Identical to AutoCAD, execute the Undo command while still placing lines. Undo is always transparent in any MicroStation command. Try using the Ctrl + Z shortcut to execute the Undo command. Continue to hold down the Ctrl key and every Z that is typed in will execute an additional Undo command.

Undo (Undo the Entire Line Segment). Identical to AutoCAD, stop the command using the Reset button (right mouse click). Use the Undo command to undo the entire line segment. Continue to repeat the Undo as often as required.

Close (Place Last Point at First Point). Snap to the first line segment start point. There is no "automatic" close option available.

AUTOCAD TIP: *Using the C shortcut out of habit will issue the AccuDraw Center Snap option. Be careful of this one!*

AutoCAD Command Comparison

AutoCAD	MicroStation
Mouse pick: Keyboard: *Line* Line	Use AccuDraw to specify precision input, or use the Tool Settings dialog to specify Length and Angle.
Close Option	There is no automatic way to "close" the line element back to the first point issued.

DRAW A CIRCLE

There are several options available for controlling how a circle is drawn. The Tool Settings dialog provides a single location for all options, which makes it easier to learn each tool. The Tool Settings dialog can be compared to the settings and options found in several locations in AutoCAD, such as the command line, the Properties dialog, and various menus (depending on the command).

Use the Tool Settings dialog with AccuDraw to minimize the steps required to complete the commands. First, draw a circle by choosing a method such as Center, Edge, or Diameter. Based on this selection, you are provided the necessary settings associated with each choice.

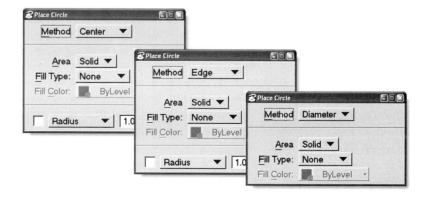

Circle by Center

1   Identify center in drawing.

2   Key in radius or diameter.

Circle by Edge

1   Identify three (3) points in the drawing to define the edge of the circle.

Circle by Diameter

1   Identify center in drawing.

2   Key in diameter.

It's that simple!

AutoCAD Command Comparison

AutoCAD	MicroStation
Mouse pick: Keyboard: *Circle* Circle	Use the Circle tool in MicroStation along with the Tool Settings dialog to find matching command options.
3P (Circle by 3 points)	Use the Edge method Tool Settings.
2P (Circle by 2 points)	Use the Diameter method Tool Settings.
TTR (Circle by Tangent-Tangent-Radius)	Use the Edge method from the Tool Settings dialog. Define the radius. Use the Tangent Snap mode.

DRAW AN ARC

There are several options for controlling how an arc is drawn. The Tool Settings dialog provides a single location for all options.

First, draw an arc by choosing a method, such as Center or Edge. Based on this selection, you are provided the necessary settings associated with each choice.

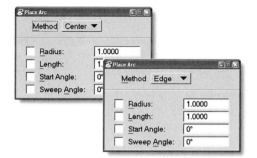

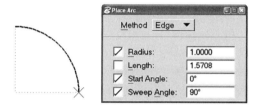

You can completely define the arc parameters using the Tool Settings dialog. The following is an example.

Radius:	1
Length:	Not defined
Start Angle:	0
Sweep Angle:	90

Arc by Center

1 Identify center location of the circle in the drawing.

Arc by Edge

1 Identify the start point location for the circle in the drawing.

AutoCAD Command Comparison

AutoCAD	MicroStation
Mouse pick: Keyboard: *Arc* Arc	Use the Arc tool in MicroStation along with the tool settings described in the following to find matching command options. MicroStation allows you to draw an arc in the clockwise and counterclockwise directions. This is a great feature for simplifying arc placement.
Arc by Start, Center, End	Define the start and end points graphically. Use the Center method from the Tool Settings dialog. Define the Radius graphically using AccuDraw.
Arc by Start, Center, Angle	Define the start and end points graphically. Use the Center method from the Tool Settings dialog. Define the Angle graphically using AccuDraw.
Arc by Start, Center, Length	Define the start and end points graphically. Use the Center method from the Tool Settings dialog. Define the Length graphically using AccuDraw.

AutoCAD	MicroStation
Arc by Start, End, Direction	Define the start and end points graphically. Define the direction graphically using AccuDraw. MicroStation allows you to draw arcs in any direction by default.
Arc by Start, End, Radius Arc by Center, Start, End Arc by Start, End, Angle Arc by Center, Start, Length Arc by Center, Start, Angle	These point orders are not available in MicroStation.

DRAW A BLOCK

A block in AutoCAD is not the same thing in MicroStation. Remember, the terminology is slightly different, and as an AutoCAD user you need to learn this new language. A block in MicroStation is equivalent to a "rectangle" in AutoCAD. To draw a rectangular element in MicroStation, use the tool settings to control the method, such as Orthogonal or Rotated. Be sure to use AccuDraw to simplify this command. The steps for creating two types of block follow.

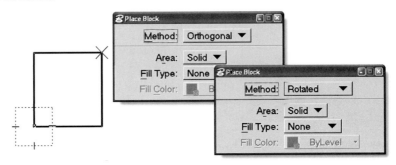

ORTHOGONAL BLOCK

1 Identify the start point of the shape.

2 Drag the cursor in one axis direction (X) and key in a distance.

3 Drag the cursor in the other axis direction (Y) and key in a distance.

4 Accept the shape size by issuing a data point anywhere in the drawing. This point will control the block direction.

ROTATED BLOCK

1 Identify the start point of the shape.

2 Drag the cursor in one axis direction (Y) and key in a distance.

3 Use the Tab key to navigate to the Angle setting in AccuDraw and key in the desired angle.

4 Accept these settings by issuing a data point anywhere in the drawing. This point will control the block direction.

5 Drag the cursor in the other axis direction (X) and key in a distance.

6 Accept this shape size by issuing a data point anywhere in the drawing. This point will control the block direction.

AutoCAD Command Comparison

AutoCAD	MicroStation
Mouse pick: Keyboard: *RECtangle* Rectangle	Use the Block tool in MicroStation along with the following Tool Settings dialog options to find matching command options.
Chamfer—Shape with chamfered vertices	Use the SmartLine tool to draw a shape with chamfered vertices. Use the Chamfered type vertex. Define the Chamfer Offset. Activate the Join Elements option.
Elevation—Shape at specified Z depth	Using a 2D file, your Z depth is locked to 0 by default. Using a 3D file, use the Set Active Depth tool along with the Depth Lock option to lock your graphics to a specific Z depth. Alternatively, use AccuDraw to define the Z value.
Fillet—Shape with filleted vertices	Use the SmartLine tool to draw a shape with filleted vertices. Use the Rounded type of vertex. Define the Rounding Radius. Activate the Join Elements option.
Thickness—Shape with Z thickness defined	Use a 3D file and extrude a Z distance from the original line element.

AutoCAD	MicroStation
Width	Use the *element weight* attribute to get the "thick" line appearance. For varied width in polylines, use a custom line style in MicroStation.
Area	There is no equivalent MicroStation command to place a shape by area.
Dimensions	Use the Block tool and AccuDraw to graphically define the shape dimensions.
Rotation	Use the Block tool and AccuDraw to graphically define the shape rotation.

In Exercise 4-6, following, you have the opportunity to practice using MicroStation's basic drawing commands for architectural application.

EXERCISE 4-6: USING BASIC DRAW COMMANDS IN AN ARCHITECTURAL EXAMPLE

In this exercise you will learn to use the basic drawing options available in MicroStation (such as SmartLine, Line, Arc Circle, and Block) in an architectural example.

1   Open the design file *DRAW_ARCH.DGN*.

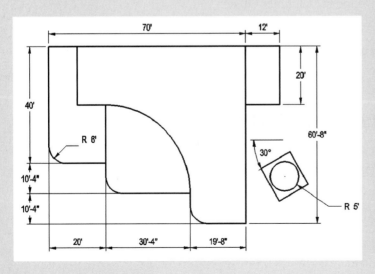

The figure at left shows a simple building layout that can be created using these simple drawing commands. The dimensions are provided for informational purposes only. You will learn to add dimensions in a later chapter.

First, we need to draw the building outline using the SmartLine tool.

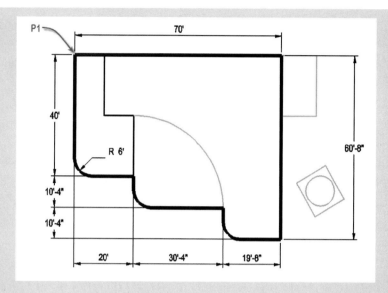

2 Select the SmartLine tool and issue a data point in view 1. This point is labeled P1 in the above figure.

3 Drag the cursor in the X direction (→) and key in the value *70 (70'-0")*. Issue a data point to accept.

4 Drag the cursor in the Y direction (↓) and key in the value *60:8 (60'-8")*. Issue a data point to accept.

5 Drag the cursor in the X direction (←) and key in the value *19:8 (19'-8")*. Issue a data point to accept.

6 Modify the following Tool Settings options.

   Vertex Type: Rounded
   Rounding Radius: 6'-0"

7 Drag the cursor in the Y direction (↑) and key in the value *10:4 (10'-4")*. Issue a data point to accept.

8 Drag the cursor in the X direction (←) and key in the value *30:4 (30'-4")*. Issue a data point to accept.

9 Drag the cursor in the Y direction (↑) and key in the value *10:4 (10'-4")*. Issue a data point to accept.

10 Drag the cursor in the Y direction (←) and key in the value *20 (20'-0")*. Issue a data point to accept.

11 Drag the cursor in the Y direction (↑) and key in the value *20 (20'-0")*. Issue a data point to accept.

12 Modify the following Tool Settings option.

       Vertex Type: Sharp

13 Drag the cursor in the Y direction (↑) and key in the value *20 (20'-0")*. Issue a data point to complete the building outline.

Next, we will add interior building division using lines and arcs.

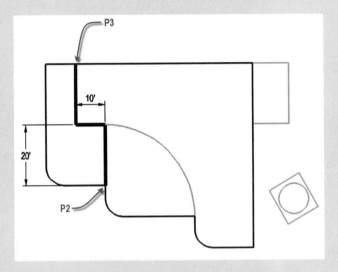

14 Select the Line tool and keypoint snap to point P2 for the start of the line segment.

15 Drag the cursor in the Y direction (↑) and key in the value *20 (20'-0")*. Issue a data point to accept.

16 Drag the cursor in the Y direction (←) and key in the value *10 (10'-0")*. Issue a data point to accept.

17 Drag the cursor in the Y direction (↑) and use the Perpendicular option to snap to point P3, shown in the previous figure.

       HINT: Use the Shift + right-click procedure to access a single-shot Perpendicular snap.

Next, we will add the arc to the building divisions.

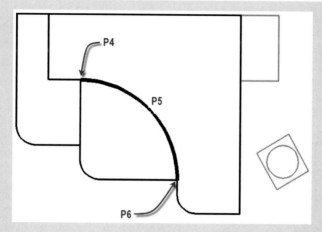

18 Select the Arc tool and verify the Tool Settings dialog's options below. Snap to P4 as the start point of the arc.

    Method: Edge
    Radius: 30'

19 Issue a data point for the arc edge midpoint in the vicinity of P5.

20 Snap to point P6 to complete the arc command.

Next, we will place a stair access attached to the upper right-hand corner of the building outline.

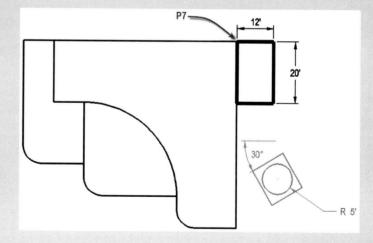

21 Select the Place Block tool and snap to the keypoint at P7 for the start the block shape.

22 Drag the cursor in the X direction (→) and key in the value *12 (12'-0"). Do not accept this point.*

23 Drag the cursor in the Y direction (↓) and key in the value *20 (20'-0")*. Now you can issue a data point to complete the block shape.

Finally, we will place a rotated block and a circle in the empty space to the right of the building.

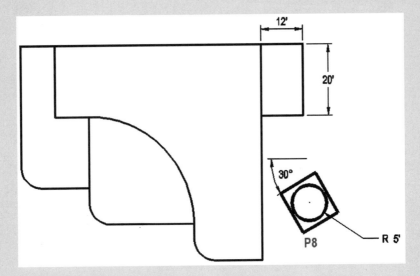

24 Select the Place Block tool and issue a data point in the vicinity of P8 for the start point of the block shape.

Set the Tool Settings method to Rotated.

25 Drag the cursor in the direction of 30 degrees (↗).

Examine the AccuDraw dialog and use the Tab key to move the focus to the Angle field. Key in the value *30* degrees to guarantee an exact angle.

26 Issue a data point to define the first segment of the rotated shape.

27 Drag the cursor in the opposite 30-degree direction (↖) and issue a data point to complete the rotated shape.

28 Select the Circle tool and use the Center Snap option to snap to the center of the rotated block.

29 Drag the cursor and key in a value of *5 (5')* for the circle radius. Issue a data point to accept and complete the circle.

In Exercise 4-7, following, you have the opportunity to practice using MicroStation's basic drawing commands in a civil engineering example.

EXERCISE 4-7: USING BASIC DRAW COMMANDS IN A CIVIL ENGINEERING EXAMPLE

In this exercise you will learn to use the basic editing options available in MicroStation (such as Copy, Copy Parallel, Scale, Rotate, and Mirror) for civil engineering application.

1   Open the design file *DRAW_CIVIL.DGN.*

The following figure shows a simple cul-de-sac layout that can be created using these simple drawing commands. The dimensions are provided for informational purposes only. You will learn to add dimensions in a later chapter.

First, we need to draw the road using the SmartLine tool and the AccuDraw shortcut option Set Origin.

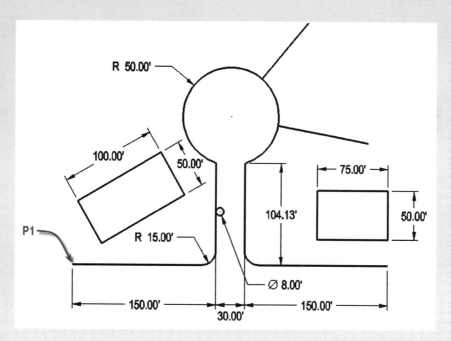

2   Select the SmartLine tool and issue a data point in view 1. This point is labeled P1 in the following figure.

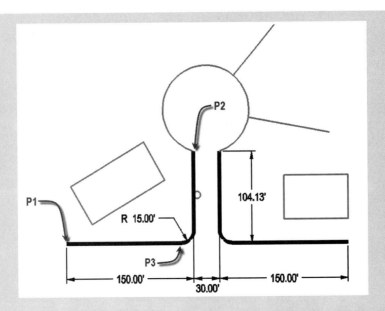

3 Drag the cursor in the X direction (→) and key in the value *150.00*. Issue a data point to accept.

4 Modify the following Tool Settings options.

   Vertex Type: Rounded
   Rounding Radius: 15.00'

5 Drag the cursor in the Y direction (↑) and key in the value *104.13*. Issue a data point to accept.

6 Use the Reset button on the mouse to end the current SmartLine operation.

7 Place a tentative point at P2 and key in the letter *O* to activate the Set Origin shortcut. You should see the AccuDraw compass move to point P2.

HINT: A tentative point is performed by clicking the middle mouse button or by clicking the right and left mouse buttons simultaneously.

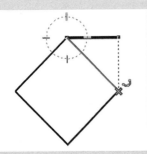

8 Modify the following Tool Settings option.

   Method: Center

9 Drag the cursor in the X direction (→) and key in the value *30.00*. Issue a data point to start the SmartLine.

10 Drag the cursor in the Y direction (↓) and press the Enter key to activate the AccuDraw SmartLock shortcut.

Move the cursor to P3 but do not pick this point immediately. Instead, "hover" over P3 until the AccuDraw Alignment indicator displays.

Only when you see this alignment indicator should you issue a data point to accept.

11 Drag the cursor in the X direction (→) and key in the value *150 (150.00')*. Issue a data point to accept.

Next, we will draw the arc for the end of the cul-de-sac.

12 Select the Arc tool, keypoint snap to point P4, and issue a data point to accept.

13 Modify the following Tool Settings options.

   Method: Edge
   Radius: 50.00'

14 Use the keypoint method to snap to point P5, and then issue a data point to accept. You might need to swing the arc around the imaginary center point to get the arc radius required.

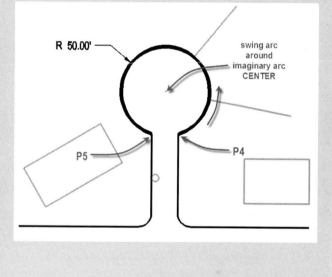

Next, we will draw the lot lines using the Perpendicular snap to guarantee they are perpendicular to the arc.

15 Select the Line tool and issue a data point in the vicinity of P6.

16 Use the Perpendicular snap option to snap to the arc at point P7 using a tentative. Issue a data point to accept.

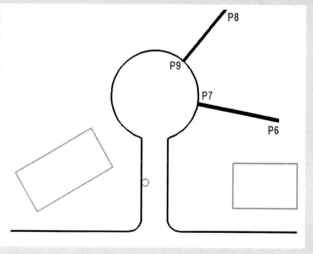

Reset to end the current command.

17 Issue a data point in the vicinity of P8.

18 Use the Perpendicular snap option to snap to the arc at point P9 using a tentative.

Reset to end the current command.

Next, we will place a power pole in the road segment using the circle command.

19 Select the Circle tool and snap to the midpoint of the vertical line at P10.

Modify the following Tool Settings option.

   Method: Diameter

20 Drag the cursor in the X direction (→) and key in the value *8.00.* Issue a data point to accept.

Next, we need to place the buildings using the Block tool.

21 Select the Place Block tool and issue a data point in the vicinity of P11 to start the block shape.

22 Modify the following Tool Settings option.

   Method: Orthogonal

23 Drag the cursor in the X direction (→) and key in the value *75.00. Do not accept this point.*

24 Drag the cursor in the Y direction (↑) and key in the value *50.00.* Issue a data point to complete the block shape.

Finally, we will place a rotated block for the second building outline.

25 Select the Place Block tool.

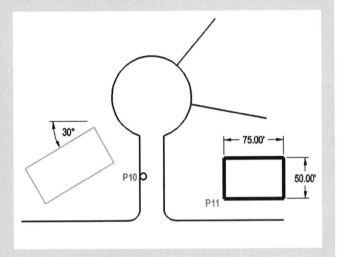

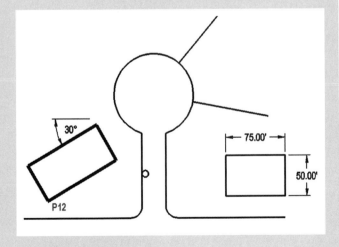

26 Modify the following Tool Settings option.

    Method: Rotated

27 Issue a data point in the vicinity of point P12 and drag the cursor in the 30-degree direction (↗).

28 Examine the AccuDraw dialog and use the Tab key to move the focus to the Angle field. Key in the value *30* degrees to guarantee an exact angle.

29 Issue a data point to define the first segment of the rotated shape.

30 Drag the cursor in the 120-degree direction (↖) and issue a data point to complete the rotated shape.

Draw a Pattern

All hatches are referred to as patterns in MicroStation. There are several types of patterns available, including hatches, cross-hatches, and linear patterns. The following command methods are available from the Tool Settings dialog for controlling the patterning area.

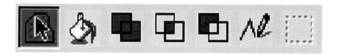

Element: Patterns an area defined by a single element.

Flood: Patterns an area defined by searching for "bounding" elements.

Union: Patterns an area defined by joining element shapes.

Intersection: Patterns an area defined by overlapping element shapes.

Difference: Patterns an area defined by the first element, using all other elements as subtracted elements.

Points: Patterns an area defined by user-specified points. No boundary element is required.

Fence: Patterns an area defined by a fence shape. No boundary element is required.

Draw a Hatch Pattern

Use the following Tool Settings dialog options to define specific pattern sizes and angles.

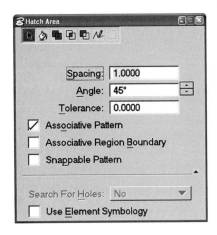

Spacing: Defines the distance between hatched lines

Angle: Defines the angle of the hatched lines

Tolerance: Defines the distance between a curved boundary element and patterned line segments

Associative Pattern. With this option, the pattern is associated with the boundary element. This causes the pattern to automatically update whenever the boundary element is modified. The pattern is automatically placed on the same level as the boundary element.

Associative Region Boundary. With this option, the pattern can be placed on a different level than the boundary element.

Snappable Pattern. This option allows you to control whether the individual pattern elements are snappable.

Search for Holes. When this option is turned on, the pattern will recognize internal "hole" element types and leave them as "holes" in the resulting pattern.

Use Element Symbology. With this option, the pattern will be placed using the boundary element symbology for color, weight, and line type.

AutoCAD Command Comparison

AutoCAD	MicroStation
Mouse pick: Keyboard: *BHatch* Hatch	Use the Hatch Area, Crosshatch Area, or Pattern Area tools
Hatch Object	Use Element method from the Tool Settings dialog. Define the spacing between the hatch lines. Define the Angle of the hatch lines. Activate the Associative Pattern option. Activate the Associative Pattern Boundary option. Activate the Snappable Pattern option.

AutoCAD	MicroStation
Hatch Object	Use the Ctrl key to remove and add boundaries in the selection set tools.
Add Pick Points	Use the Flood Area tool from the Tool Settings dialog to flood an area with the selected hatch or pattern.
Add Select Objects	Use the Element Area tool to select elements to be hatched. To hatch more than one element, use a selection tool such as the Element Selection or Power Selector to select multiple elements prior to executing the Hatch Area command.
Remove Boundaries	There is no way to remove a boundary from an existing hatched element. You can use the Delete Pattern tool to delete a specific pattern from an element or use points to define a non-associated hatch without a boundary.
Recreate Boundary	Using Region Associative Boundary will allow you to disassociate a pattern to boundary elements but remain intelligent enough to repair as needed. A disassociated pattern will display in alternate symbology using a very heavy and dashed pattern. Once the boundary elements are repaired, the associated pattern will revert to the correct symbology.
Create Separate Hatches	MicroStation always creates separate patterns for each element selected.
Draw Order	Draw order is not controlled from within the pattern commands. A specific Draw Order command is not available in MicroStation V8.
Inherit Properties	Use the Change Properties tool to modify existing hatch settings from one hatch pattern to match another. Use the Match/Change method. Use Smart Match or Match Pattern to set hatch settings prior to creating a new hatch pattern.

AutoCAD	MicroStation
Use Current Origin and Specified Origin	MicroStation does not use a default base point for the default hatch origin. The origin of the pattern is controlled by the boundary element selection point. You can modify the origin point by issuing a tentative point during the Accept step of the pattern command. Select the boundary element. Tentative snap to the preferred origin point. Data point to accept the new origin point and complete the pattern command. Use the Change Pattern tool to modify the hatch origin of an existing hatch. Activate the Intersection Point option.

DRAW A CROSS-HATCH PATTERN
Use the additional Tool Settings dialog options to define specific pattern sizes and angles.

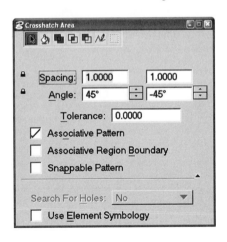

Spacing: Defines the distances in both directions for the cross-hatch lines.

Angle: Defines the angles for both cross-hatch lines.

Tolerance: Defines the distance between curved boundary elements and the pattern line segments.

DRAW A SYMBOL PATTERN
A symbol pattern allows you to create a repeating graphical cell that can be used to fill an area quickly and easily.

Pattern Definition: Defines the type of repeatable graphic pattern to be used. You can choose a cell from any available cell library. Use the Browse Cell button to navigate to a specific cell library to locate a cell.

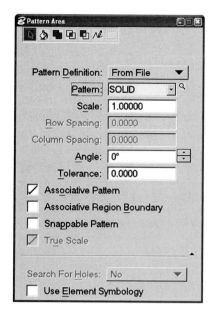

You can select a pattern from an AutoCAD *.PAT* file using the From File pattern definition option. By default, MicroStation will automatically locate any existing AutoCAD installation *.PAT* files. Use the *Browse for .PAT File* button to navigate to a specific AutoCAD *.PAT* file. Select the desired hatch pattern from the list provided.

Scale: Defines the scale of the repeated pattern cell or file.

Row Spacing: Defines the horizontal distance between the symbol pattern.

Column Spacing: Defines the vertical distance between the symbol pattern.

True Scale: Enables the automatic scaling of cell or pattern resource files with varying working unit definitions.

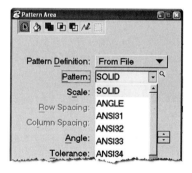

DRAW A LINEAR PATTERN

A linear pattern allows you to modify the appearance of a linear element without a custom line type. This command places a repeated pattern end to end along any linear element based on the preferred cycle type.

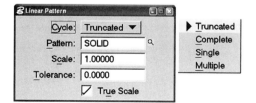

Truncated: Slices off any remaining pattern from the last instance at the end of the linear element.

Complete: Forces the entire pattern to be placed on the linear element, resulting in a varied pattern scale from one linear element to another.

Single: Forces a single pattern onto the linear element, resulting in a stretched or compressed pattern scale from one linear element to another.

Multiple: Forces a complete pattern cycle at the end of the linear element unless the remaining length of the linear element is less than 80%

of the pattern instance. This option may result in a varied pattern scale from one linear element to another.

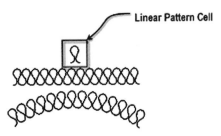

MicroStation uses cells to define linear patterns, which is very simple and easy—especially if you have ever tried to make a custom line style in AutoCAD. For example, the batting insulation cell displayed in example 1 at left shows how simple the graphic cell is. The cell need only contain a single pattern cycle. Likewise with examples 2 and 3 shown below.

Example 1: Batting Insulation Linear Pattern

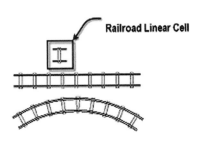

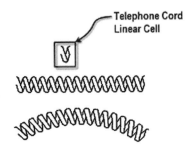

Example 2: Railroad Linear Pattern

Example 3: Telephone Cord Linear Pattern

DRAW A SOLID FILL

A solid fill is not a separate element in MicroStation, and thus cannot be accomplished with the hatch or pattern commands. A solid fill is considered a "property" of a closed-shape element (similar to weight, color, or line type). This property is available from the Place Shape commands via the Tool Settings dialog.

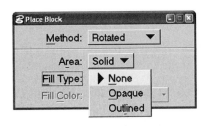

For example, when placing a shape element using the Place Block tool you can set the Fill Type to None, Opaque, or Outlined. An example of this output is shown below.

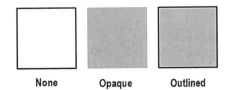

None   Opaque   Outlined

None: No solid fill is applied to the shape.

Opaque: The solid fill is applied with no boundary edge.

Outlined: The solid fill is applied with a boundary edge.

5: Basic Element Editing

CHAPTER OBJECTIVES:

❑ Learn to use basic manipulation and modification tools

❑ Learn the basics of selection tools

❑ Learn how to manipulate and modify multiple elements

BASIC ELEMENT EDITING

This chapter introduces basic editing functions, such as copy, move, scale, rotate, and so on. In this chapter we focus on how to use these editing tools to modify and produce drawings. Learn to take advantage of your AutoCAD skills and apply them in using MicroStation efficiently. In the sections that follow, you will learn to use the Tool Settings dialog to access common command options.

The Basics

COPY ELEMENT
Use the Copy tool to replicate existing content throughout a drawing. The options available with this command allow for multiple copies simultaneously. In the Tool Settings dialog, key in the number of copies required and specify the copy direction using AccuDraw. The copy is repeated by the number of copies.

AutoCAD Command Comparison

AutoCAD	MicroStation
Mouse pick: Keyboard: *COpy* Copy With the addition of the Multiple copy option in AutoCAD 2006, these commands are functionally identical.	Use the Number of Copies tool setting to save repetitive copy steps. Use AccuDraw to simulate the direct distance entry and polar manipulation.
Basepoint	Identical functionality using data point.
Displacement	Identical functionality using data point.

MOVE ELEMENT

Use the Move tool to relocate existing content throughout the drawing. The options available with this tool allow for multiple copies simultaneously. If the Copies option is activated, the command is converted into the Copy tool (discussed previously).

AutoCAD Command Comparison

AutoCAD	MicroStation
Mouse pick: Keyboard: *Move* Move These commands are functionally identical.	Use the Number of Copies tool setting to save repetitive move steps. Use AccuDraw to simulate the direct distance entry and polar manipulation.
Basepoint	Identical functionality using data point.
Displacement	Identical functionality using data point.

MOVE/COPY PARALLEL

Use the Move Parallel option to offset existing elements in the drawing. This tool is commonly used to copy parallel rather than move parallel. The modes available determine how SmartLine vertices are handled.

Mitre Vertex. Extends or shortens angles connected with matching vertices.

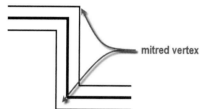

mitred vertex

1 Select the Mitre method.

2 Activate the distance setting and key in a distance.

3 Select the Make Copy setting if you want to copy parallel.

4 Select the element to be copied.

5 Pick the side to be offset.

You can issue additional data points to continue copying parallel lines.

Round Vertex. This method forces outside vertices to fillet automatically. The radius of the fillet is determined by the offset distance.

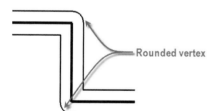

Rounded vertex

1 Select the Round method.

2 Activate the distance setting and key in a distance.

3 Select the Make Copy setting if you want to copy parallel.

4 Select the element to be copied.

5 Pick the side to be offset.

You can issue additional data points to continue copying parallel lines.

Original Vertex. Using this method forces outside vertices to maintain the vertex type of the original element.

1 Select the Original method.

2 Activate the distance setting and key in a distance.

3 Select the Make Copy setting if you want to copy parallel.

4 Select the element to be copied.

5 Pick the side to be offset.

Define Distance Graphically . Use the Define Distance button to define the distance graphically when using the Move/Copy Parallel tool.

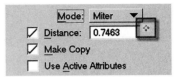

Use Active Attributes. This setting causes the new offset element to use the active attribute settings for level, color, line style, and line weight.

AutoCAD Command Comparison

AutoCAD	MicroStation
Mouse pick: Keyboard: *Offset* Offset	These commands are functionally identical.
Distance	Use the Mitre Method from the Tool Settings dialog. Key in the offset distance or use the Define Distance button to graphically define the distance. Activate the Make Copy option to access the Copy Parallel tool.
Through	Use the Mitre Method from the Tool Settings dialog. Key in the offset distance or use the Define Distance button to graphically define the distance. Activate the Make Copy option to access the Copy Parallel tool.
Erase	Use the Mitre Method from the Tool Settings dialog. Key in the offset distance or use the Define Distance button to graphically define the distance. Turn off the Make Copy option to access the Move Parallel tool.
Layer	Use the Mitre Method from the Tool Settings dialog. Key in the offset distance or use the Define Distance button to graphically define the distance. Activate the Make Copy option to access the Copy Parallel tool. Activate the Use Active Attributes option use the active level, color, line style, and line weight.

SCALE ELEMENT

Use the Scale tool to increase or decrease the size of elements in a drawing. There are several options available for controlling how elements are scaled.

Scale by Active Scale

1 Key in the scale factor.

2 Identify the element to be scaled.

3 Identify the Origin point to scale about.

The active scale is proportional when the "padlock" icon is locked. You can unlock this icon by clicking on it. When unlocked, the X and Y scales can differ, which provides a nonproportional scaling factor.

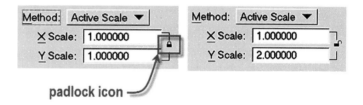

padlock icon

TIP: *The Active Scale fields are calculator friendly, and you can compute scale factors directly in the X, Y, and Z scale fields. For example, you can enter the ratio 125/75 and MicroStation converts it to 1.667. If you watch closely you can see the calculator appear in the Tool Settings dialog when you key in the mathematical expression.*

Scale by 3 Points

1 Identify the elements to be scaled.

2 Identify the Origin point to scale about.

3 Identify the existing scale reference point.

4 Identify the new scale reference point.

Proportional. This setting is available only during the 3 Point scale method. Activating this setting will guarantee a proportional scale result.

Copies. This setting allows you to scale a copy of the original elements, leaving the original elements intact.

About Element Center. This setting allows you to control the scale origin point without having to define it. This is especially useful when scaling multiple elements in a selection set. MicroStation picks the centroid (center of mass) to scale about. You can scale open or closed elements about their centers.

AUTOCAD TIP: *When using the Tool Settings dialog, be sure to check for the Additional Settings icon. This icon should be activated to display any additional settings available for the active command. Many times the most productive settings are "hidden" from view when this icon is not activated.*

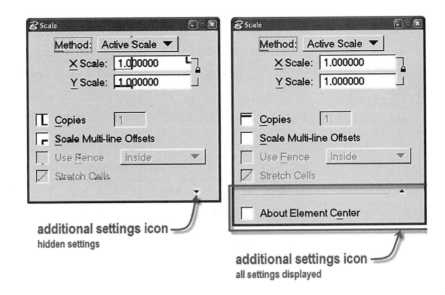

additional settings icon — hidden settings

additional settings icon — all settings displayed

AutoCAD Command Comparison

AutoCAD	MicroStation
Mouse pick: Keyboard: [icon] *SCale* Scale	These commands are functionally identical.

AutoCAD	MicroStation
Scale a Copy	Use the Active Scale Method from the Tool Settings dialog. Activate the Copies option. Specify the number of copies required.
Scale by Reference	Use the 3 Points Scale Method from the Tool Settings dialog. Activate the proportional option.

ROTATE ELEMENT

Use the Rotate option to modify the angle of an existing element. There are several options available for controlling how elements are rotated.

Rotate by Active Angle

1 Key in the active angle.

2 Identify the element to be rotated.

3 Identify the Origin point to rotate about.

Rotate by 2 Points

1 Identify the element to be rotated.

2 Identify the Origin point to rotate about.

3 Graphically specify the rotation angle. Be sure to use AccuDraw to simplify this method.

Rotate by 3 Points

1 Identify the elements to be rotated.

2 Identify the Origin point to rotate about.

3 Identify the existing rotation reference point.

4 Identify the new rotation reference point.

Copies. This setting allows you to rotate a copy of the original elements, leaving the original elements intact.

About Element Center. This setting allows you to control the rotation origin point without having to define it. This is especially useful when rotating multiple elements in a selection set. MicroStation picks the centroid (center of mass) to rotate about. You can rotate open or closed elements about their centers.

AutoCAD Command Comparison

AutoCAD	MicroStation
Mouse pick: Keyboard: *ROtate* Rotate	
Rotate a Copy	Use the Active Scale Rotate Method from the Tool Settings dialog. Activate the copies option. Specify the number of copies required.
Rotate by Reference	Use the 3 Point Scale Method from the Tool Settings dialog.

MIRROR ELEMENT

Use the Mirror tool to flip existing objects about an axis. There are several options available for controlling how the elements are mirrored.

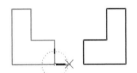

Mirror Horizontal

1   Select the elements to be mirrored.

2   Identify the horizontal axis location.

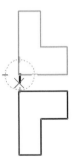

Mirror Vertical

1   Select the elements to be mirrored.

2   Identify the vertical axis location.

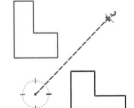

Mirror About Line

1   Select the elements to be mirrored.

2   Identify the first point on the mirror axis.

3   Identify the second point on the mirror axis.

Make Copy.. This setting allows you to mirror a copy of the original elements, leaving the original elements intact.

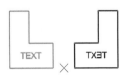

Mirror Text. This setting allows you to mirror text elements if needed. This option is useful if you need a reverse image of a drawing.

AutoCAD Command Comparison

AutoCAD	MicroStation
Mouse pick: Keyboard: *Mirror* Mirror The AutoCAD command does not distinguish between horizontal, vertical and about a line axis definitions. This requires that you must define two points for every mirror.	Use the Mirror tool to access similar functionality in MicroStation.
Mirror a Copy	Use the Horizontal, Vertical or Line Mirror Methods from the Tool Settings dialog. Activate the copies option. Specify the number of copies required.
Erase Source Objects?	Use the Make Copy setting to control how the original elements are managed during a mirror operation.
The **MIRRTEXT** system variable in AutoCAD controls how text is handled during a mirror operation.	Activate the Mirror Text setting.

In Exercise 5-1, following, you have the opportunity to practice performing basic editing commands.

EXERCISE 5-1: BASIC EDITING COMMANDS

In this exercise you will learn to use the basic editing functions available in MicroStation. These include Copy, Copy Parallel, Scale, Rotate, and Mirror.

1 Open the design file *BASIC_ EDIT1.DGN*.

In the first few steps we will copy the existing desk object to begin the furniture layout.

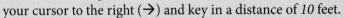
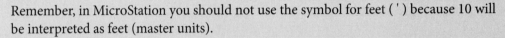

2 Select the Copy tool.

3 Select the desk in view 1 and, using AccuDraw, drag your cursor to the right (→) and key in a distance of *10* feet.

Remember, in MicroStation you should not use the symbol for feet (') because 10 will be interpreted as feet (master units).

4 Issue a data point to accept the new location.

5 Undo this change.

That simple copy was easy enough, but we need more than one desk in the layout.

6 Select the Copy tool and set the number of copies to 3.

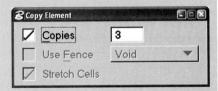

7 Select the desk in view 1 and, using AccuDraw, drag your cursor to the right (→) and key in a distance of *10* feet.

8 Issue a data point to accept the new location.

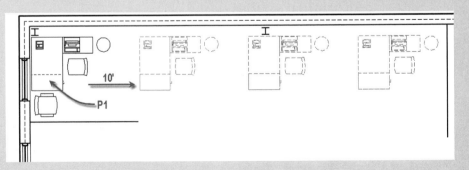

In the next few steps you will learn to use the Copy Parallel tool to finish the horizontal wall for the cubicle. We want to parallel copy this line at the same wall thickness as the vertical wall already completed.

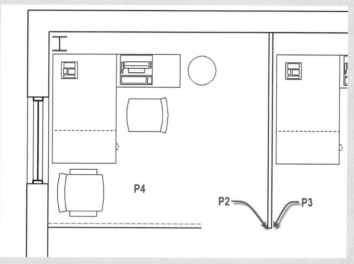

9 Select the Copy Parallel tool and activate the Make Copy setting.

10 Click on the Define Distance button and pick the vertical wall distance at P2 and P3. Verify that your distance setting is set to 0:2 (2 inches).

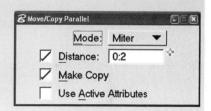

11 Select the horizontal wall and issue a data point at P4 to copy parallel above the existing line. Be sure to issue a reset (right mouse button) so that you don't accidentally copy another parallel line.

12 Using the Line tool, finish the end cap of the new wall.

Additional Practice: More Copy Parallel
Add the remaining stair treads to the right side of the stair. This tread distance should be equal to the left side of the stair. Be sure to use the multiple capabilities of the Copy Parallel tool to increase your productivity on this task.

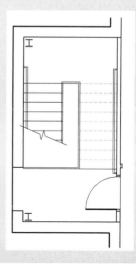

ALIGN ELEMENT EDGES

Use the Align tool to line up existing elements to the edge of another element. This align tool allows you to align a single element or multiple elements along any of the following orthogonal axes

❑ Top	❑ Bottom
❑ Left	❑ Right
❑ Horizontal	❑ Center
❑ Vertical center	❑ Both centers

Align Top

1   Set the Align Tool Settings option to TOP.

2   Select the element to be aligned to.

3   Select elements to be aligned.

Align Bottom

1   Set the Align tool to BOTTOM.

2   Select the element to be aligned to.

3   Select elements to be aligned.

Align Left

1   Set the Align tool to LEFT.

2   Select the element to be aligned to.

3   Select elements to be aligned.

Align Right

1   Set the Align tool to RIGHT.

2   Select the element to be aligned to.

3   Select elements to be aligned.

Align Horizontal Center

1 Set the Align tool to HORIZONTAL CENTER.

2 Select the element to be aligned to.

3 Select elements to be aligned.

Align Vertical Center

1 Set the Align tool to VERTICAL CENTER.

2 Select the element to be aligned to.

3 Select elements to be aligned.

Align Both Centers

1 Set the Align tool to BOTH CENTERS.

2 Select the element to be aligned to.

3 Select elements to be aligned.

AutoCAD Command Comparison

AutoCAD	MicroStation
Keyboard: *ALign* The Align tool is buried in the 3D operations of AutoCAD and many of you probably didn't even know it was there. It requires that you define two points on the objects to be aligned and two points to define the alignment axis.	

AutoCAD	MicroStation
Align Source Point, Align Destination Point	If you used the Align objects tool to align to a none orthogonal axis the only way to do this in MicroStation is to use the Move and Rotate tools individually.
Align Source Point, Align Destination Point, Scale objects	If you used the Align objects tool to scale the objects during the alignment operation you must use the MicroStation Scale tool independently of the alignment operation.

ARRAY ELEMENTS

Use the Array tool to copy elements in an arranged pattern. This pattern can be in a rectangular shape defined by rows and columns, or in a polar shape arranged in a circular pattern.

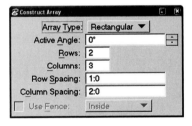

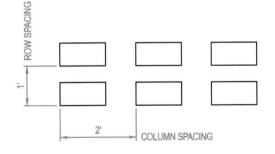

Rectangular Array

1   Set the Array Type option to Rectangular.

2   Define the Active Angle of the array.

3   Define the number of Rows in the array.

4   Define the number of Columns in the array.

5   Define the Row Spacing.

6   Define the Column Spacing.

Polar Array

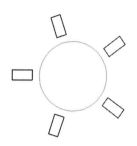

1 Set the Array Type option to Polar.

2 Define the number of Items in the array.

3 Define the Delta Angle between the items in the array.

4 Activate the Rotate Items setting.

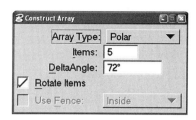

AutoCAD Command Comparison

AutoCAD	MicroStation
Mouse pick: Keyboard: *ARray* Array	The array tools are very similar with a few exceptions. The settings for methods, alternate settings for angle, and the array preview differ slightly in MicroStation.
Rectangular	The settings for a rectangular array are identical to those in AutoCAD. Use the Rectangular Array Type. Define the Active Angle. **NOTE:** *There is no way to graphically define the angle.* Define the number of Rows. Define the number of Columns. Define the Row Spacing. **NOTE:** *There is no way to graphically define the row distance.* Define the Column Spacing. **NOTE:** *There is no way to graphically define the column distance.*
Polar	The only AutoCAD method available in MicroStation is: Total number of items and angle between items.

AutoCAD	MicroStation
Total number of items and angle to be filled. Angle to be filled and angle between items.	These AutoCAD methods are not available in MicroStation: The angle setting for a polar array is a delta angle between items, not the entire polar angle to be filled.
	The ability to rotate items is available in MicroStation. However the ability to define a pre-determined base point is not available. This is probably not a problem for most of you, in that rarely does an AutoCAD user take advantage of this option anyway. Use the Polar Array Type. Define the number of items. Define the Delta Angle between items. **NOTE:** *There is no way to graphically define the delta angle.* Activate the Rotate Items option. **NOTE:** *There is no way to graphically define the item rotation base point.*

In Exercise 5-2, following, you have the opportunity to practice using the Align command.

EXERCISE 5-2: USING THE ALIGN COMMAND

In this exercise you will learn to use the Align tool to clean up the window tags in this messy drawing.

1   Open the design file *ALIGN1.DGN*.

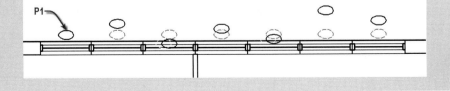

2 Select the Align tool and set the alignment setting to Horizontal Center.

3 Select the ellipse at P1 as the element to be aligned to.

4 Identify each ellipse that needs to realigned with this correct window tag.

5 Close the design file *ALIGN1.DGN*.

Element Selection Tools

The element selection tools provided in MicroStation are very powerful and flexible. With these tools you can use your AutoCAD skills and learn some of the new options available only in MicroStation. Remember that you must select elements before selecting commands.

ELEMENT SELECTOR
This tool is the fundamental selection tool used to select single or multiple elements for easy manipulation.

Single Element Selection. To select a single element, click on the element to identify it and then select the applicable manipulation tool.

Multiple Elements Selection. To select multiple elements, drag a selection box around the elements to be selected. The direction of this selection box does not affect the selection mode. This type of selection box will only select elements completely inside the box.

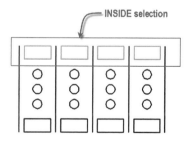

To select elements both inside and overlapping the selection box, use the Ctrl + Shift keys while dragging the selection box around the elements to be selected. The Ctrl + Shift keys activate the overlap mode of the selection tool.

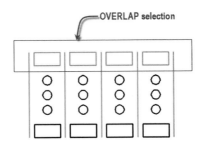

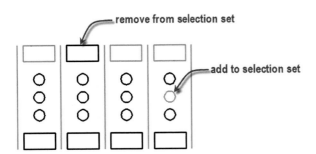

AUTOCAD TIP: *There is no physical difference in the appearance of the "in-side only" selection box and the "inside and overlapping" selection box. The use of the Ctrl + Shift keys determine this selection mode change.*

You can add and remove elements to the selection set individually using the Ctrl key. Hold down the Ctrl key while picking elements to be added to or removed from a selection set.

De-Select All. To clear (cancel) a selection set, issue a data point anywhere in the view window where there are no selectable elements.

AUTOCAD TIP: *Using the Esc key to clear the selection set will do nothing. There is no way to force the Esc key to clear a selection set. Sorry!*

POWER SELECTOR

The PowerSelector most closely matches the selection tools found in AutoCAD. Using this tool provides detailed feedback and tool settings that make the creation of complex selection sets much easier.

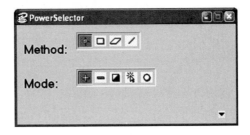

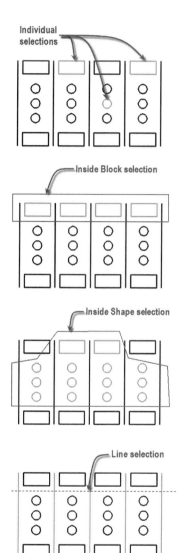

Individual selections

Inside Block selection

Inside Shape selection

Line selection

Selection Methods. There are four selection methods available using PowerSelector.

Individual: Allows you to select elements individually based on the active selection mode.

Block (Inside): Allows you to select elements using a rectangular shape based on the active selection mode. Only elements completely inside will be selected. This is identical to the window selection method in AutoCAD.

Shape (Inside): Allows you to select elements using a nonrectangular shape based on the active selection mode. Only elements completely inside will be selected. This is identical to the window polygon selection method in AutoCAD.

Line (Overlap): Allows you to select elements using a line shape based on the active selection mode. Only elements overlapping the line will be selected. This is identical to the fence selection method in AutoCAD.

Two other selection options (Block and Shape) are not initially apparent. To access these additional tools, you must click on the associated icons a second time to toggle their method. The icon will change in appearance to represent which method (Inside or Overlap) is active.

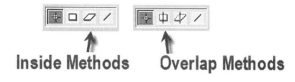

Inside Methods   **Overlap Methods**

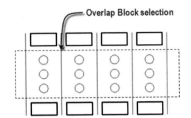

Overlap Block selection

Block (Overlap): Allows you to select elements using a rectangular shape based on the active selection mode. Only elements completely inside and overlapping will be selected. This is identical to the crossing selection method in AutoCAD.

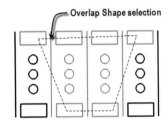

Overlap Shape selection

Shape (Overlap): Allows you to select elements using a nonrectangular shape based on the active selection mode. Only elements completely inside and overlapping will be selected. This is identical to the crossing polygon selection method in AutoCAD.

Selection Modes. There are five selection modes available using PowerSelector.

 Add: Allows you to add elements to the selection set.

 Subtract: Allows you to remove elements from the selection set.

 Invert: Allows you to toggle the selection status of elements. Elements already selected are deselected, and elements that are not selected are selected.

 New: This option clears the current selection set and defines a new selection set in a single command operation. The first point of your selection method actually issues the Clear command.

 Clear/Select All: Allows you to clear the active selection set, or to select all elements in a file if no selection set is active.

AUTOCAD TIP: *Don't forget to check for more settings by using the "additional settings" icon. Remember, some of the most productive settings are "hidden" here.*

 The "Hidden" Options. When you expand the "additional settings" icon you will discover several lists of element attributes relating to the active selection set and drawing. Using these lists you can filter your selection set to just what you need, especially if the elements you want to manipulate are not easily selected graphically.

For example, you can add elements to the selection set graphically, and then you can remove all of the text.

Level (LV Tab): Allows you to modify the content of a selection set based on level names or numbers.

Color (CO Tab): Allows you to modify the content of a selection set based on color.

Line Style (LC Tab): Allows you to modify the content of a selection set based on line style.

AUTOCAD TIP: The LC abbreviation on the tab is a throwback to the early days of MicroStation, when line style was called "linecode."

Weight (WT Tab): Allows you to modify the content of a selection set based on line weight.

Element Type (TY Tab): Allows you to modify the content of a selection set based on element type, such as text, dimensions, lines, arcs, and so on.

Class (CL Tab): Allows you to modify the content of a selection set based on element class. Refer to the section "Element Classifications" in Chapter 4 for more detailed information on element class.

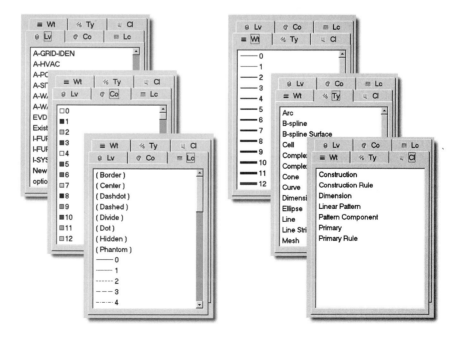

AutoCAD Command Comparison

AutoCAD uses basically two methods to select objects: a window and a crossing. These methods are determined by the direction in which the user drags the cursor across the drawing window

AutoCAD	MicroStation
Window Selection This selection method is activated by dragging the cursor left to right, and is visually represented by a solid line and a blue fill type.	Use the PowerSelect tool. Select the Block method and verify that the Inside icon is displayed. Select the Add mode.
Crossing Selection This selection method is activated by dragging the cursor right to left and is visually represented by a dashed line and a green fill type.	Use the PowerSelect tool. Select the Block method and verify that the Overlap icon is displayed. Select the Add mode.
Fence This selection method is activated by drawing a line through elements in the drawing and is visually represented by a dashed line.	Use the PowerSelect tool. Select the Line method. Select the Add mode.
Window Polygon This selection method is activated by the key-in *WP* and visually represented by a solid line and a blue fill type.	Use the PowerSelect tool. Select the Shape method and verify that the Inside icon is displayed. Select the Add mode.
Crossing Polygon This selection method is activated by the key-in *CP* and visually represented by a dashed line and a green fill type.	Use the PowerSelect tool. Select the Shape method and verify that the Overlap icon is displayed. Select the Add mode.
Add to Selection Set By default, AutoCAD always adds objects to an existing selection set. There are two methods to accomplish this in MicroStation.	Use the PowerSelect tool. Select the appropriate selection method Select the Add mode. Hold down the Ctrl key to add elements to an existing selection set.

AutoCAD	MicroStation
Remove from Selection Set In AutoCAD you must hold down the Shift key to remove objects from an exiting selection set. There are two methods to accomplish this in MicroStation.	Use the PowerSelect tool. Select the appropriate selection method. Select the Subtract mode. 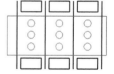 Hold down the Ctrl key to remove elements from an existing selection set.
Clear Selection Set AutoCAD uses the Esc key to clear all selection sets. There are two methods to accomplish this in MicroStation.	Use the PowerSelect tool. Select the Clear mode. 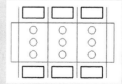 Issue a data point (left mouse button) anywhere in the view window where there are no elements.

In Exercise 5-3, following, you have the opportunity to practice using the Selection tools.

EXERCISE 5-3: USING THE SELECTION TOOLS

In this exercise you will learn to use the default Selection tool to manipulate elements.

1 Open the design file *SELECTION1.DGN*.

In the next few steps you will learn to select multiple elements and make global modifications. This exercise is set up in a puzzle fashion to make sure you can select the required elements. Have fun!

Puzzle 1

The Default Selection Tool

First, select just the circles.

1 Select the Selection tool and drag a shape around the circles. That was easy!

Second, select the circles and the lines at the same time.

2 Select the Selection tool and hold down the Ctrl + Shift keys while dragging a box around the circles. Not too bad, right?

Third, select the lines and the rectangles. You will either need to remove some elements from the selection set or perform two selection operations.

3   Select the Selection tool and drag a box around all of the graphics.

4   Hold down the Ctrl key and drag a box around the circles to remove them from the current selection set.

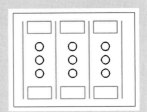

The PowerSelector

First, select just the circles.

5   Select the PowerSelector tool and use the Block (Inside) method and the Add mode.

6   Place a selection shape around the circles.

Clear the current selection set.

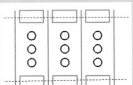

Second, select the circles and the lines at the same time.

7   Select the PowerSelector tool and use the Block (Overlap) method and the Add mode.

8   Place a selection shape around the circles.

Clear the current selection set.

Third, select the lines and the rectangles. You will need to draw two lines to get everything.

9   Select the PowerSelector tool and use the Line method and the Add mode.

10 Place a selection line through the top rectangles and the bottom rectangles.

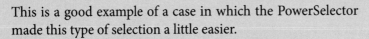

This is a good example of a case in which the PowerSelector made this type of selection a little easier.

Additional Practice: More Selection Sets

Puzzle 2: Selection Commands

Try the following.

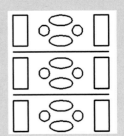

❑   Select the lines only.

❑   Select the ellipses only.

❑   Select the circles only.

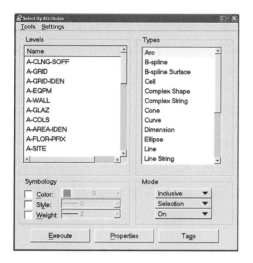

SELECT BY ATTRIBUTES

There is still another way to create complex selection sets that cannot be created graphically. The Select By Attributes utility provides you with a dialog-driven selection tool based on several element symbologies, types, properties, and content. Use this utility to make global changes to your drawing based on specific selection criteria.

For example, you need to plot a drawing for a presentation so you need to modify the font type and text size.

Another example: you need to replace an existing cell with a new cell symbol throughout the entire drawing. Check this tool out. Its power is quite amazing.

AutoCAD Command Comparison

AutoCAD	MicroStation
Quick Select This AutoCAD tool allows you to select objects by properties rather than graphically.	To obtain the same functionality in MicroStation use the Select By Attributes tool.
	Upon exiting this tool you are asked the following question. **Alert** (!) WARNING: OK will allow SELECTBY to continue to filter location or display of elements. CANCEL will stop element filtering by SELECTBY. OK Cancel Don't worry most of us didn't understand it the first time either! Basically either answer is correct when you are working with selection sets. The only differences apply when using Select By for other purposes.

THE HISTORY OF FENCES

The Fence tool has traveled through time and has a lot of history in the DGN world. Long ago, before anyone even thought about selection sets, a tool called Fence provided a method of gathering up elements for manipulation. This tool is still widely used, by us older MicroStation users, but in my opinion the use of selection sets and the tools discussed earlier are much more powerful and intuitive than the Fence tool. However, because you may sit next to one of "us" (old MicroStation users) and hear about placing fences (and because a "fence" in AutoCAD is something very different), let's examine the functionality the Fence tool provides.

What Is a Fence?

A fence is primarily used for selecting and clipping things. A fence is a "screen" element that does not really exist except on the screen. It isn't a normal element like a line or an arc that can be deleted. It is somewhat permanent when placed, in that the only way to get rid of a fence is to select the Place Fence tool again or to draw an Element Selection window.

Fence Pros and Cons

The following outline the pros and cons of using fences.

PROS

❑ Familiar to old users

❑ Best method for stretching elements

❑ Good method for clip-masking reference files

❑ Good method for clipping raster files

❑ Is a temporary element

CONS

❑ Does not highlight elements for visual clarification of element set

❑ Cannot dynamically see elements during manipulation commands

❑ Cannot filter elements using level, color, line style, line weight, or element type

❑ Can only select elements using "area" shapes

Using Fences

The sections that follow describe various fence operations.

PLACE FENCE

Use the Fence tool to select multiple elements for manipulation. There are basically two types of settings when using a fence: the type of fence and the mode used to select elements.

Copy Using a Fence

1   Select the Place Fence tool.

2   Select the fence type.

3   Select the fence mode.

4   Place the fence shape in the view window.

5   Select an element manipulation tool.

6   Activate the Use Fence setting, which instructs MicroStation to use the fence shape for this manipulation.

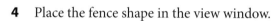

After you activate the Use Fence setting, you can change the fence mode if needed.

Fence Type. Defines what the fence looks like. The following fence type options are available.

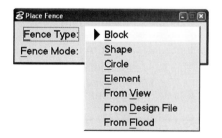

Block: Allows you to select elements for manipulation using a rectangular shape based on the active fence mode.

Shape: Allows you to select elements for manipulation using a nonrectangular shape based on the active fence mode.

Circle: Allows you to select elements for manipulation using a circular shape based on the active fence mode.

Element: Allows you to select elements for manipulation using an existing "element" shape based on the active fence mode.

From View: Allows you to select elements for manipulation using a view window based on the location of a data point.

From Design File: Allows you to select elements for manipulation using design file extents based on the active fence mode.

From Flood: Allows you to select elements for manipulation using a shape generated with a flood fill based on the location of a data point.

Fence Mode. Defines how the fence works. Note the icon displayed in the status bar reflecting the active fence mode.

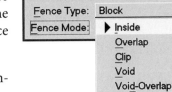

 Inside: Select elements completely inside the fence shape.

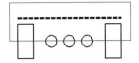

Inside Fence

 Overlap: Select elements inside and overlapping the fence shape.

 Clip: Select elements inside the fence shape, and break any elements that overlap the fence shape.

 Void: Select elements completely outside the fence shape.

 Void Overlap: Select elements outside and overlapping the fence shape.

 Void Clip: Select elements outside the fence shape, and break any elements that overlap the fence shape.

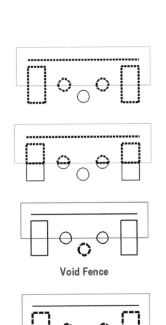

Void Fence

Void Overlap Fence

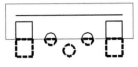

Void Clip Fence

STRETCH COMMAND

Use the Fence Stretch tool to lengthen elements in the drawing.

Stretch Using a Fence

1 Select the Place Fence command.

2 Select the Fence Type needed.

3 Select the Fence Mode needed.

4 Place the fence shape in the view window.

5 Select the Manipulate Fence Contents command.

6 Set the fence Operation setting to Stretch.

7 Turn on the Stretch Cells setting if needed.

Stretch Cells. Use this setting if you need to stretch cell symbols. By default, cells do not stretch like other element types.

AutoCAD Command Comparison

AutoCAD	MicroStation
Mouse pick: Keyboard: *Stretch* Stretch	AutoCAD uses a crossing window to stretch objects, and Micro-Station requires the use of a fence to stretch elements. However, MicroStation does not care what the fence mode is, so feel free to use any mode to stretch elements.
Stretch an Element	Use the Manipulate Fence tool. Set the Operation setting to Stretch. Select the applicable fence Mode. Activate the Stretch Cells option if needed.

In Exercise 5-4, following, you have the opportunity to practice using the fence functionality.

EXERCISE 5-4: USING A FENCE

In this exercise you will learn to use the MicroStation Fence tool to stretch elements.

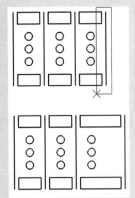

1   Open the design file *STRETCH1.DGN*.

In the next few steps you will stretch and move the rectangles and the line located on the right-hand side of the graphics at left.

2   Select the Place Fence tool and place a fence block around the line and through the rectangles, as shown in the previous figure at left.

3   Select the Fence Manipulation tool and set the operation type to Stretch.

4   Using AccuDraw, identify a point on one of the line end points, drag the cursor to the right, and key in a distance of *1*.

5   Select the Place Fence tool again to remove the fence.

6   Close the design file *STRETCH1.DGN*.

BEYOND THE BASICS

More Element Editing

This section explores some additional editing commands found in Micro-Station. You will find that these commands are some of the most used commands on a daily basis. We all spend more time editing than actual drawing, and thus the efficiency of these edits is critical to our productivity.

MODIFY ELEMENT
Use the Modify Element tool to modify any element type. The capabilities of this tool vary based on the type of element selected. In fact, it is one of the most capable and powerful editing tools in MicroStation.

If you use this tool on a linear element, you can modify the end points, segment length, and vertices. When used on a radial element, you can modify the radius or diameter.

The type of modification is also dependent on where you select the element. If you select a linear element at a vertex, you can modify that vertex location. If you select a linear element on a segment, you can modify that segment location.

If you modify a dimension element, you can move existing components (such as the extension line, dimension line, or text) to new locations. If you modify a SmartLine, you can change the actual vertex geometry and construction.

PARTIAL DELETE

Use the Partial Delete tool to remove just a portion of an element. You can remove a portion of an element between any two points along the element itself

AutoCAD Command Comparison

AutoCAD	MicroStation
Mouse pick: Keyboard: *BReak* Break at Point	Use the Partial Delete tool to break an element at a single point.
First Point	Select the element initially at the point where the break should start.
Second Point	Select the same point again to break the element at a single point.
Mouse pick: Keyboard: *BReak* Break	Use the Partial Delete tool to break an element between two points.
First Point	Select the element initially at the point where the break should start.
Second Point	Select the element at the point where the break should end. **NOTE:** *Arcs will only break in a counterclockwise direction.*

EXTEND LINE

Use the Extend Line tool to extend or shorten any linear or arc element type. You can specify the distance as a positive or negative value (using this distance to add or subtract from the existing length). You can also specify the length of the element from the nearest end point. The end point closest to your selection point will be modified.

1   Select the element to extend.

2   Drag the element to its new end-point location.

Distance. Specify a positive distance to lengthen the element, or a negative distance to shorten it.

From End. Use the From End setting to specify that the distance be calculated from the origin point on the line regardless of where it is selected.

AutoCAD Command Comparison

AutoCAD	MicroStation
Keyboard: *LENgthen*	Use the Extend Line tool to lengthen an element.
Delta	Use the Distance tool setting to define the change in length.
Percent	Use the Scale tool to lengthen using a scale percentage.
Total	Use the From End setting to key in the total length.
	Use the Extend Line tool and AccuDraw using the O shortcut key to define the total length of an element.
Dynamic	Use the Extend Line tool with no tool settings defined to lengthen an element graphically.
	Use AccuDraw to assist with this type of lengthen.

EXTEND ELEMENTS TO INTERSECTION

Use the Extend Elements to Intersection tool to extend two elements to their intersecting point. Both elements' lengths are modified using this command, and the lengths can be lengthened or shortened. You should select the elements on the segments you wish to keep.

AutoCAD Command Comparison

AutoCAD	MicroStation
Mouse pick: Keyboard: ![fillet icon] *FILLET* Fillet with radius = 0	Use the Extend Elements to Intersection tool to modify two elements to a clean corner.
Polyline	Use the SmartLine tool to draw the original element to get these features. Using a SmartLine will allow you to modify one or more vertices of the element.
Radius	Use the Fillet tool to specify a radius of the vertex cleanup.
Trim	Use the Fillet tool and the Truncate cleanup option to specify the type of cleanup required: None, Both, or First.
Multiple	All MicroStation commands are multiple.
Undo	Use Ctrl + Z to perform an undo from within the command. Use the Undo button to execute an undo within the command.

EXTEND ELEMENT TO INTERSECTION

Use the Extend Element to Intersection tool to extend a single element to the intersection of another element. Only the element selected is modified, and the end point closest to your selection point is lengthened or shortened.

AutoCAD Command Comparison

AutoCAD	MicroStation
Mouse pick: Keyboard: ![extend icon] *EXtend* Extend	Use the Extend Element to Intersection tool to lengthen an element to another element.
Fence	Use the PowerSelector tool with the Line method to select elements.
Crossing	Use the PowerSelector tool with the Block or Shape method in overlap mode to select elements.

AutoCAD	MicroStation
Project	All elements can extend to edges that exist and those that are projected in 2D and 3D space.
Edge	All elements can extend to edges that exist and those that are projected in 2D and 3D space.
Undo	Use Ctrl + Z to perform an undo from within the command. Use the Undo button to execute an undo within the command.
Shift + Pick to Trim	Use the Intelli-Trim tool to perform both trim and extend operations within the same command.

TRIM
Use the Trim tool to cut and shorten existing elements using single or multiple elements in the drawing as cutting edges or boundaries. This command works best when using a selection set, but can be used by selecting individual elements.

INTELLI-TRIM
Use this "intelligent" trim tool for additional functionality and cutting capabilities not found in the previous trim tool. You can use the Intelli-Trim tool to cut, extend, or trim elements with each other.

Quick. The Quick tool setting allows you to modify the basic functionality of the command from a shortening and lengthening modification to a simple cutting operation. Use the Trim, Extend, or Cut tool to determine which modification type you prefer.

Advanced. The Advanced tool setting toggles the tool between the trimming and extending modes. Once the mode has been defined, you can change the order of the selection process using the options Select Elements to Trim or Select Elements to Extend.

AutoCAD Command Comparison

AutoCAD	MicroStation
Mouse pick: Keyboard: TRim Trim	Use the Trim or Intelli-Trim tools along with selection sets to access identical functionality in MicroStation.
Fence	Use the PowerSelector tool with the Line method to select elements.
Crossing	Use the PowerSelector tool with the Block or Shape overlap method to select elements.
Project	All elements can extend to edges that exist and those that are projected in 2D and 3D space.
Erase	Use the Delete tool after completing the trim operation.
Undo	Use Ctrl + Z to perform an undo from within the command. Use the Undo button to execute an undo within the command.
Shift + Pick to Extend	Use the Intelli-Trim tool to perform both trim and extend operations within the same command.

INSERT VERTEX

Use the Insert Vertex tool to add vertices in an existing element. You can control the physical location of the new vertex using snap modes.

1   Identify the element at the location to add the vertex.

2   Issue a data point to accept and locate the new vertex.

DELETE VERTEX

Use the Delete Vertex tool to remove vertices from an existing element.

1   Identify the element at the vertex to be removed.

2   Issue a data point to accept.

> **TIP:** *You can use the Insert Vertex and Delete Vertex tools to add and remove extension lines in dimensions.*

AutoCAD Command Comparison

AutoCAD	MicroStation
Mouse pick: Keyboard: *PEdit* Pedit	Use the Insert or Delete Vertex tool to add and remove vertices from shapes and continuous line elements.
Edit Vertex	Not required. Pick the vertex to be inserted or removed.
Next	Not required.
Previous	Not required.
Straighten	Not required.
Go	Not required.
Exit	Not required.

FILLET ELEMENTS

Use the Fillet Elements tool to round existing vertices on elements.

1 Select the first element.

2 Select the second element.

Radius. Use the Radius setting to specify the radius of the new vertex.

Truncate. Use the Truncate setting to specify how the existing elements should be modified. The options available are None, Both, or First. See the figure at left for examples of these truncation options.

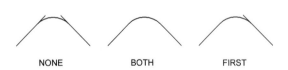

AutoCAD Command Comparison

AutoCAD		MicroStation
Mouse pick: Fillet	Keyboard: *FILLET*	Use the Fillet tool to modify the intersection of two elements with a rounded corner.
Polyline		Use the SmartLine tool to draw the original element to get these features. Using a SmartLine will allow you to modify one or more vertices of the element.
Radius		Use the Radius tool setting to specify a radius for the rounded corner.
Trim		Use the Truncate cleanup option to specify the type of cleanup required: None, Both, or First.
Multiple		All MicroStation commands are multiple.
Undo		Use Ctrl + Z to perform an undo from within the command. Use the Undo button to execute an undo within the command.

CHAMFER ELEMENTS

Use the Chamfer Elements tool to connect two elements using a specific angled line. The angle of the line is calculated from the existing element's angle.

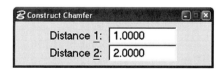

Distance 1. Use the Distance 1 setting to specify the distance along the first element selected to begin the new angled vertex.

Distance 2. Use the Distance 2 setting to specify the distance along the second element selected to end the new angled vertex.

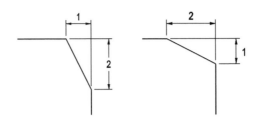

AutoCAD Command Comparison

AutoCAD	MicroStation
Mouse pick:　　Keyboard:　*CHAMFER*　　Chamfer	Use the Chamfer tool to modify the intersection of two elements with an angled corner.
Polyline	Use the SmartLine tool to draw the original element to get these features.　Using a SmartLine will allow you to modify one or more vertices of the element.
Distance	Use the Distance 1 and Distance 2 tool settings to define the distance along the two elements of the corner.
Angle	Not available.
Trim	Use the Truncate cleanup option to specify the type of cleanup required: None, Both, or First.
Method	Not available.
Multiple	All MicroStation commands are multiple.
Undo	Use Ctrl + Z to perform an undo from within the command.　Use the Undo button to execute an undo within the command.

In Exercise 5-5 following, you have the opportunity to practice editing elements.

EXERCISE 5-5: MORE EDITING

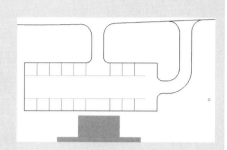

In this exercise you will learn to use several common modification commands with existing elements. Here you will use the Modify, Partial Delete, Extend Line, Trim, Insert Vertex, Delete Vertex, and Fillet tools.

1 Open the design file *MODIFY2.DGN*.

The next few steps will walk you through the procedure for modifying the parking lot and the parking access roads.

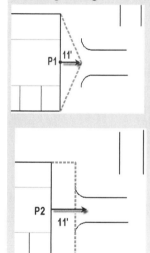

2 Select the Insert Vertex tool to add a vertex at the midpoint at P1.

Drag the cursor in the X direction (→) and key in the distance of *11* feet.

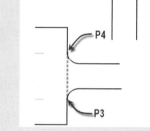

3 Select the Delete Vertex tool and remove the vertex you just added.

4 Select the Modify tool and add 11 feet to the east end of the parking lot shape.

Select the midpoint at P2 and using AccuDraw drag the cursor in the X direction (→) and key in the value *11* feet.

5 Select the Partial Delete tool and delete the parking lot shape between P3 and P4.

Snap to the arc end points to start and end the partial delete. You might need to use the Reset button on the mouse to access the shape (not the arc).

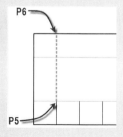

6 Select the Extend Element to Intersection tool to extend the parking stripe from P5 to P6.

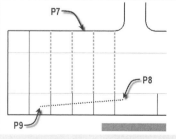

7 Select the parking stripe, and then select the horizontal line of the parking lot to extend to.

8 Select the Intelli-Trim tool and establish the following tool settings.

> *Mode:*   Quick
> *Operation:* Extend

9 Select the line to be extended to by selecting the parking lot boundary at P7.

The boundary line element should change to a "dashed symbology" when selected as the boundary edge.

10 Define the "line" selection method by defining a crossing line from P8 to P9.

The stripe lines should automatically extend to the parking lot boundary.

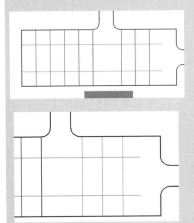

11 Repeat the Intelli-Trim command to extend the remaining parking stripe lines to the boundary shape.

12 Use the Selection tool to select the lines highlighted in the below left figure.

13 Select the Trim tool and pick the parking stripe lines on the portion you want to remove (noted as P10 through P12 in the figure below).

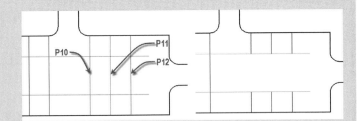

14 Select the Intelli-Trim tool and establish the following tool settings.

> *Mode:*   Advanced
> *Operation:*   Trim
> Select Cutting Elements

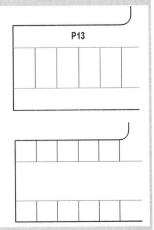

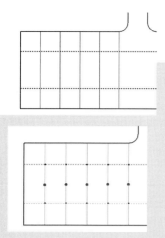

15 Select the horizontal lines as the cutting elements.

16 Issue a Reset button on the mouse and select the vertical lines at the midpoints designated in the figure at left.

The intersection "dots" are created when you select the vertical lines (as shown below left).

17 Right-click and the trim will preview.

18 The trim may be reversed from what you were expecting, but this is just a preview. You can toggle it to the other direction.

19 Move the cursor to the outside of the cutting edges in the area designated by P13 and issue a data point.

This should reverse the trim again, and if this trim is correct reset to complete the command.

In Exercise 5-6 following, you have the opportunity to practice modifying lines.

EXERCISE 5-6: MODIFY MORE LINES

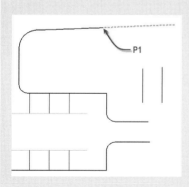

1 Select the Extend Line tool and set the distance tool setting to *50* feet.

2 Select the right-of-way line at P1 to extend that end of the line.

If you want to control the total length of the line, use the following steps with the Extend Line tool. In this example we will change the total length of this line to *100* feet.

3 Select the Extend Line tool and deactivate all tool settings.

4 Select the right-of-way line and when using AccuDraw the compass should locate itself at the west end point of this line.

HINT: If the compass is not at the westmost end point of this line, use the O shortcut key to relocate it.

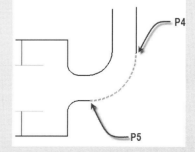

Verify that you are using the "polar" (round) compass.

5 Key in the distance value of *100* (feet).

Issue a data point to accept this value.

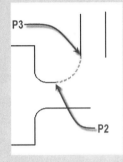

6 To complete the exit lanes from the parking lot, select the Fillet tool and establish the following tool settings.

> *Radius:* 11 (feet)
> *Truncate:* Both

7 Select the Fillet tool and establish the following tool settings.

> *Radius:* 21 (feet)
> *Truncate:* Both

8 Select the lines at P4 and P5 to complete the command.

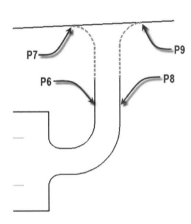

9 Select the Fillet tool and establish the following tool settings.

> *Radius:* 10 (feet)
> *Truncate:* First

10 Select the lines at P6 and P7 to extend the line, and add the fillet to the first line selected. Note that the second line is not modified.

11 Repeat the Fillet command and select the lines at P8 and P9.

6: Controlling Drawings

CHAPTER OBJECTIVES:

❑ Learn about element display properties

❑ Learn about levels and level symbology

❑ Learn to use reference files

❑ Learn to use raster files

The objective of this chapter is to teach you how to use some of the basic concepts associated with CAD and the DGN file format. Discover how you can use levels and level symbology to reduce editing requirements and to produce flexible output. Learn how to more fully utilize other CAD data to your advantage.

CONTROLLING DRAWINGS

There are several aspects of the DGN file that allow you to control what data you see and how that data appears. These are levels, color, line weight, and line styles, discussed in the sections that follow.

Level Management and Display

One of the most powerful enhancements found in MicroStation V8 is the implementation of unlimited levels. Previous versions of MicroStation limited the level structure to 63 levels. With this in mind, let's explore how DGN CAD standards were developed and how they work within this environment.

THE OLD 63 LEVELS
Prior to V8, and with only 63 levels to work with, if you wanted to distinguish drawing components from each other you used element attributes such as color, line weights, and line styles to make different aspects of the

drawing unique. It was common to have multiple colors, line weights, and even line styles on a single level. In many legacy DGN drawings you will find a single level containing continuous and dashed lines, red and blue lines, or thin and thick lines. This was common practice and considered "normal" in the MicroStation DGN file.

This is very different from the way things work in AutoCAD. Because AutoCAD allows you to create up 65,000+ layers, the most logical way to distinguish drawing components was through layers. If you needed continuous and dashed components, you placed them on different layers. Red and blue objects each had a unique layer, and the same held true for varying line styles. You primarily used layers with colors, line weights, and line styles directly assigned. With the introduction of V8, both of these configurations are possible in MicroStation. You can choose which method you prefer and work within that configuration.

MULTIPLE ATTRIBUTES ON A SINGLE LEVEL

In this book I will refer to the "old V7" method as the ByElement method since attributes were assigned directly to the element.

Pros. Using multiple attributes on a single level minimizes the number of levels you have to define in a single file. In nonstandard CAD environments, this method is useful because you can use any attribute you want on any level. This provides a more open drawing environment.

Cons. Using multiple attributes on a single level forces you to define all attribute types each time you place a component. You have to define color, line weight, and line style individually before you can start to compose the drawing. This method is more difficult to manage and usually requires more customizations to enforce. In large organizations, it does not lend itself to CAD standards compliance without significant effort on the part of the CAD manager or the end user.

SINGLE ATTRIBUTES ON MULTIPLE LEVELS

In this book I will refer to the "AutoCAD" and "V8" method as the ByLevel method.

Pros. Using levels as the distinguishing feature between components allows for easier management of CAD standards. All attributes can be defined to a specific level so that the user only has to select the correct level and the attribute settings will automatically follow the CAD standard for color, line weight, and line style. This system is easier to implement and tends to encourage compliance with CAD standards.

Cons. Using levels as the distinguishing feature dramatically increases the number of levels required to define a CAD standard. However, this method can restrict your ability to define attributes independently.

USING BYELEMENT

The concept of ByElement symbology is not new to the MicroStation environment, but is new to most AutoCAD users. The definition of *ByElement* is that a component's attributes are determined by the active setting at the time of placement. Each component in a drawing has a color, line style, and line weight assigned to it. If a component is placed on a level that is assigned the color white and the active color is blue, the resulting component will be blue. The color assigned to the level has no influence on the component being placed. The color of a component is controlled *by* the *element*.

USING BYLEVEL

The concept of ByLevel symbology is new to the MicroStation V8 environment, but not to AutoCAD users who have been using ByLayer symbology for several years. The definition of *ByLevel* is that a component's attribute is determined by the setting of a level, not by its individual setting. If a component is placed on a level that is assigned the color green and the active color is set to ByLevel, the resulting component will be green. If the same component is placed on a level that is assigned the color blue, the resulting component will be blue. The color of a component is controlled *by* the *level*. ByLevel allows you to modify the color of a level and have all components using that level automatically update to the new color.

USING BYCELL

The concept of ByCell symbology is also new to MicroStation V8, but again not to AutoCAD users who may have been using ByBlock symbology for several years. The definition of *ByCell* is that a component's attribute is determined by the active setting at the time of placement. If a component is assigned the color ByCell, is placed on a level that is assigned the color green and the active color is set to ByLevel, the resulting component will be green. If this same component is placed on a level that is assigned the color green but the active color is blue, the resulting component will be blue.

ByCell works exactly like ByLevel if the active attribute setting is ByLevel. If the active attribute setting is anything other than ByLevel, the component will use the active setting definition. ByCell allows you to have a "tweak factor" for color, line weight, and line style when needed.

The following comparison table outlines differences between AutoCAD and MicroStation regarding handling of level settings.

AutoCAD Command Comparison

AutoCAD	MicroStation
BYLAYER	Use the BYLEVEL level attribute setting.
BYBLOCK	Use the BYCELL level attribute setting.

Level Manager

Managing levels is a simple but tedious job at best. The good news, however, is that you only have to do it once. The bad news is that you have to do it once. The best suggestion is to define these levels in a seed file and use this seed file to create every drawing from this point forward. The levels can be defined directly in the seed file or in a DGNLIB style file that is attached to the seed file.

The advantage of creating the levels in a DGNLIB file, commonly referred to as a style library file, is that it is a single source of levels that can reside in a common location on a server. This DGNLIB file can be "linked" or "embedded" into the seed file depending on your corporate requirements. This DGNLIB is then a single source of levels that is easier to maintain long term. The location of these DGNLIB files is managed using the configuration variable *MS_DGNLIBLIST*. The *MS_DGNLIBLIST* variable represents a list of directories to be used to store resource libraries for levels, text, dimension styles, and even multi-lines. You must decide which level definition methods you prefer (ByElement or ByLevel) and create the appropriate levels using the Level Manager dialog in either seed files or DGNLIB files.

Table 6-1 outlines an example of MicroStation's Level Manager level definitions for the ByElement, ByLevel, and ByCell methods.

TABLE 6-1: LEVEL MANAGER DIALOG ENTRIES FOR LEVEL METHODS

Method	Name	Number	Color	Line Style	Line Weight
ByElement	A-WALL	1	0	0 (CONTINUOUS)	0
ByLevel	A-WALL	2	CYAN	0 (CONTINUOUS)	1 (.18)
ByCell	A-WALL	3	CYAN	0 (CONTINUOUS)	1 (.18)

NUMBERS OR NAMES

The use of logical level names has been available for several years in Micro-Station but most users were so familiar with level numbers they never bothered to assign names. Fortunately, it appears that many are taking advantage of logical level names in V8 and thus this situation should improve in the future.

All V8 levels will be assigned logical names when migrated from V7 to V8. However, not all organizations managed this as well as they could have. You will probably run into legacy DGN files that have not been assigned logical level names and the names were generated by MicroStation automatically. These level names will appear as *Level 1*, *Level 2*, *Level 3*, and so on. You can rename these numbered names to something more logical using a remapping file.

All unused levels should have been removed during the migration process from J to V8. If not, you can delete unused levels manually using the key-in

Level Delete levelname. The names of used levels are displayed in bold. The active level is displayed in an aqua color.

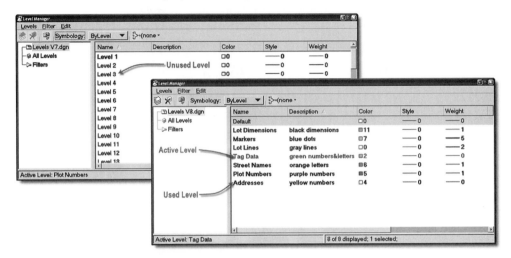

Level Manager Dialog

This section describes the options available via the Level Manager dialog. You can access this dialog by selecting **Settings** > **Levels** > **Manager** or by selecting the Level Manager button located on the Primary Tools toolbar.

New Level

The New Level option allows you to create a new level in the active design file.

> **Tip:** *The ability to create levels can be restricted using level configuration variables.*

Delete Level

The Delete Level option allows you to delete an unused level. You cannot delete a level that has been used. To view what types of elements are using a level, select the level in the list and right-click to access that level's Properties command. Select the Usage tab and a list is provided informing you what type of elements and how many are using this level. This ability to modify level properties can be restricted using level configuration variables. You may also use the key-in level usage *<level name>* to list a specific level's use. The results are reported in the Message Center.

UPDATE LEVELS

The Update Levels option allows you to update the level list from an external style library file (DGNLIB). To find out which style library is being used, select the level in the list and right-click to access that level's Properties command. Select the General tab to view the style library file name.

SYMBOLOGY

The Symbology option allows you to modify the appearance of elements in the design file. This is similar to VISRETAIN in AutoCAD but with many more capabilities. Symbology can be applied to active file levels and reference file levels. Level symbology can be toggled on and off as needed on a per-level or per-file basis. Use the tree view in the Level Manager dialog to control what levels you see in the list.

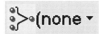

LIST FILTER

The List Filter option allows you to filter what levels are displayed in the Level Manager dialog. Using a level filter limits the level list to what you need. Typical level filter configurations might be annotation levels, survey levels, floor plan levels, used levels, unused levels, and so on. This option is invaluable when you employ the ByLevel method because long level lists can be counterproductive.

> **TIP:** *Selections made in the Level Manager dialog for List Filter will also affect levels displayed in other level interface tools, such as the Level Display dialog and the Attributes toolbar.*

MODEL LIST VIEW

The Model List View option allows you to control which model levels you want to view in the Level Manager dialog: active file models, reference file models, or both.

ALL LEVELS

The All Levels option allows you to view the levels from all models in the Level Manager dialog; both active file models and all reference file models.

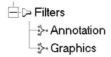

FILTERS

The Filters option allows you to view only those levels that meet specific level filter criteria. These criteria can be based on level name, level group, logical file, number, used, library, color, line weight, line style, and even other filters.

AutoCAD Command Comparison

AutoCAD	MicroStation
Mouse pick: Keyboard: *LAYER* Layer Manager	Use the Level Manager dialog to manage and create levels.
?	Select **Level Manager** > **Levels** > **Export** to view and export a level list to an external file.
Make	Use the New Level command from the Level Manager dialog. In either the Level Manager or the Level Display dialog, double-click on a level name to make it the active level.
Set	In either the Level Manager or the Level Display dialog, double-click on a level name to make it the active level.
New	Use the New Level command from the Level Manager dialog.
ON	Use the Level Display dialog or the Attribute toolbar to turn on levels.
OFF	Use the Level Display dialog or the Attribute toolbar to turn off levels.
Color	Use the Level Manager dialog to define level colors.
Ltype	Use the Level Manager dialog to define level line styles.
LWeight	Use the Level Manager dialog to define level line weights.
Plot	Use the Level Manager dialog to define whether or not a level plots.
Freeze	Use the Level Display dialog or the Attribute toolbar to turn off levels. Use the Level Display or Level Manager dialog to globally freeze levels.
Thaw	Use the Level Display dialog or the Attribute toolbar to turn on levels. Use the Level Display or Level Manager dialog to globally thaw levels.
Lock	Use the Level Display dialog or the Attribute toolbar to lock levels.
Unlock	Use the Level Display dialog or the Attribute toolbar to unlock levels.

AutoCAD	MicroStation
State	Use level symbology overrides to modify the level state found in the Level Manager dialog. There is no method for saving level states in MicroStation.

Level Display Dialog

This section describes the options found in the Level Display dialog.

VIEW INDEX

The View Index buttons allow you to control which view window levels are displayed in the level list and to control which view window(s) the level changes will be applied to.

DISPLAY TYPE

The Display Type option allows you to control which type of level display you want to use.

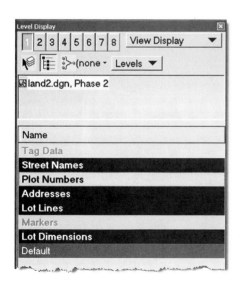

View Display: Displays levels that are turned ON/OFF in a specific view window. This is similar to the Level ON/OFF found in previous versions of Micro-Station and AutoCAD's viewport freeze and thaw.

Global Display: Displays levels that are turned ON/OFF in all view windows. Note that with this display type the view index is disabled, restricting you from specifying a specific view number. This is similar to the layer ON/OFF found in AutoCAD's model space.

Global Freeze: Displays levels that are either FROZEN (off) or THAWED (on), similar to AutoCAD's FREEZE/THAW layers. Note that with this display type the view index is disabled, restricting you from specifying a specific view number. This setting is identical in functionality to Global Display in the MicroStation environment.

The Global Display and Global Freeze options are both necessary in MicroStation when working in DWG workmode. Because AutoCAD uses both display types, and because they have different functionality in Auto-CAD, MicroStation must be able to manipulate both display types when in DWG workmode.

CHANGE LEVEL

The Change Level command allows you to modify the display of a level, its locked status, and its target.

Display Only: Changes the display of a view window to the level selected only. This works in the same manner as MicroStation's All Except Element command, and is equivalent to the Isolate Layer command found in AutoCAD's Express tools.

Display Off: Changes the levels displayed in a view window by turning off the level of a selected element. This works in the same manner as MicroStation's Off By Element command and is equivalent to the Layer Off command found in AutoCAD's Express tools.

Lock/Unlock: Changes the lock status of a level in all view windows. Locking a level using this method is different from using the Level Lock command found in previous versions of MicroStation. This new command locks the level selected so that all data on that level cannot be modified.

TIP: *The Level Lock command is exactly the opposite in functionality from this Lock Level. Level Lock restricts your access to just the level selected and all other levels cannot be modified.*

Set Target: Sets the Target master or reference file levels in Level Display of the selected element. By clicking on an element you can set the file in the target tree and see the levels for that file only.

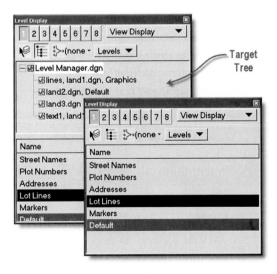

Set Active
All On
All Off
Invert Selection
Off By Element
All Except Element
Save Filter
Level Manager
Properties

Place the cursor over the level list in the Level Display dialog to access the following commands from a right-click menu.

All On: Turns on all levels in a view window.

All Off: Turns off all levels except the active level in a view window.

Off by Element: Turns off levels in a view window by selecting the elements.

All Except Element: Turns off all levels in a view window except the selected element's level.

SHOW TARGET TREE

The Show Target Tree button allows you to see the model/file list tree view in the Level Display dialog. Selecting files from this list controls which model levels you want to view in the Level Display dialog: active file, reference files, or both. You can select multiple files using the Ctrl and Shift keys.

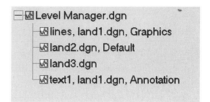

LIST FILTER

The List Filter button allows you to establish which levels are displayed in the Level Display dialog. Using a level filter shortens the level list display to what you need. Typical level filter configurations could be annotation levels, survey levels, floor plan levels, and so on. These filter settings are invaluable when using the ByLevel method because long lists of levels can be difficult to use.

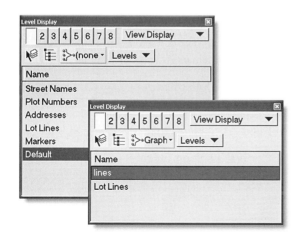

TIP: *Selections made in the Level Manager dialog for List Filter will also affect levels displayed in other level interface controls such as the Level Display dialog and the Attributes toolbar.*

LEVELS OR FILTERS VIEW

The Levels or Filters View button allows you to view individual levels associated with the models, or level filters associated with the models. Using level filters is an efficient way of turning multiple levels on and off quickly. Using an annotation level filter, for example, you could turn off all annotation with a single click.

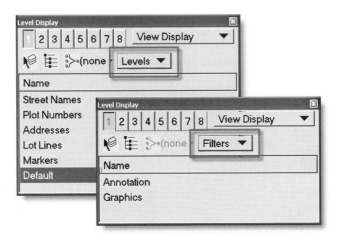

LEVEL DISPLAY LIST

The Level Display section allows you to manipulate which levels are on or off. The content of this list is controlled by the various settings and selections in the top portion of the dialog. So what do all of those colors mean?

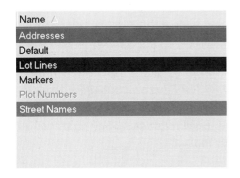

AQUA: Represents the active level

BLACK: Represents a level that is ON

GRAY: Represents a level that is OFF

DARK GRAY: Represents a level whose on/off status varies between the views or models selected

DISABLED GRAY: Represents a level that is globally off or globally frozen

BOLD: Represents a level that contains elements

NON-BOLD: Represents an empty level

The Attribute Toolbar

Levels can also be controlled using the Attribute toolbar. This toolbar is most likely where you will make common changes to the active level, color, line style, and line weight. You can use this toolbar to control not only the active attributes but whether levels are on/off or locked/unlocked.

Active Attribute
Display Window

ACTIVE LEVEL
The active level setting displays the current level for placing elements.

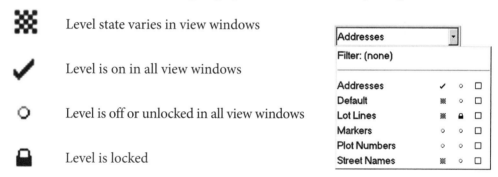

Level state varies in view windows

Level is on in all view windows

Level is off or unlocked in all view windows

Level is locked

AutoCAD Command Comparison

AutoCAD		MicroStation	
	Layer ON	✔	Toggle the Attribute tool
	Layer OFF	○	Toggle the Attribute tool
	Layer Thaw	✔	Toggle the Attribute tool
	Layer Freeze	○	Toggle the Attribute tool

AutoCAD	MicroStation
Layer LOCK	Toggle the Attribute tool
Layer UNLOCK	Toggle the Attribute tool

Element Appearance

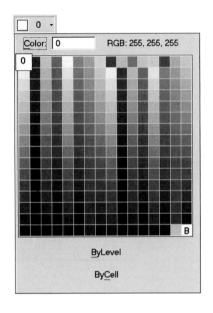

ACTIVE COLOR

The active color setting displays the active color for placing elements from the attached color table. You can access the attached color table by selecting **Settings > Color Table**.

 Represents the view window background color as color 255.

 ByLevel attribute definition. This option is available by default.

 ByCell attribute definition. This option is not available by default.

TIP: *The ByCell setting is not available by default. You can enable it through the workmode capability settings using the following configuration variable.*

```
_USTN_CAPABILITY < +CAPABILITY_BYCELL
```

You can remove the availability of the ByLevel functionality using the following workmode configuration variable.

```
_USTN_CAPABILITY < -CAPABILITY_BYLEVEL
```

Refer to the file WORKMODE.CFG for additional workmode configuration settings. This file can be found in the default installation folder C:\Program Files\Bentley\Program\MicroStation\config\system\ workmode. cfg.

ACTIVE COLOR TABLE

MicroStation allows you to attach and modify the color table as needed. You can even attach an AutoCAD color table if you prefer that color/number arrangement.

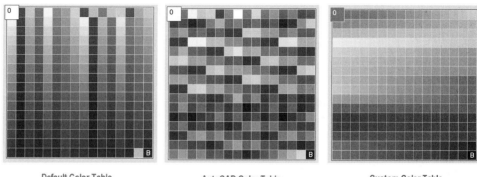

Default Color Table   AutoCAD Color Table   Custom Color Table

ACTIVE LINE STYLE

The Active Line Style setting displays the current line style for placing elements.

Line Styles 0 Through 7. The default line styles 0 through 7 are view independent, which means that if you zoom in or out their on-screen appearance does not change. They are defined using "pixels" on the screen, not physical lengths. Their plotting appearance, however, is controlled using a plot definition file that defines the line style segments in physical lengths.

Custom Line Styles. Custom line styles such as Border and Center are not view independent and their on-screen appearance will change as you zoom in and out. These line styles are similar to those found in AutoCAD. They are WYSIWYG, an acronym for "What you see is what you get." The screen appearance is exactly what you see on paper.

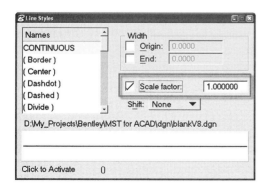

It is important to manage the scale of custom line styles. The line style definition can define a base scale, and you can modify this scale during placement. The line style scale is hardcoded to the element. You can also define the custom line style with a scale of 1 and globally modify all custom line styles using the following key-in command.

```
ACTIVE LINESTYLESCALE scalefactor
```

This command works exactly like *LTSCALE* in AutoCAD.

AutoCAD Line Styles

You can import the AutoCAD line styles from the software-delivered file *ACAD.LIN* or any custom AutoCAD *.LIN* file.

Use the MicroStation Line Style Editor dialog to create your own line styles, or to import them from another file. You can access the Line Style Editor dialog by selecting **Element > Line Styles > Edit**.

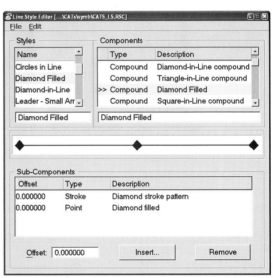

Active Line Weight

The Active Line Weight setting displays the current line weight for placing elements.

Line Weights 0-31. The default line weights 0 through 31 are view independent, which means that if you zoom in or out their on-screen appearance does not change. They are defined using "pixels" on the screen, not physical width. Their plotting appearance, however, is controlled using a plot definition file that defines the line thickness in physical widths.

This is a very efficient way of using line weight for complex CAD drawings. As an AutoCAD user, if you tried to use the new line weights you know that a hardcoded weight definition is difficult to view, and that using true "WYSIWYG" line weights is difficult to control on the screen. There are several viewing problems to deal with, and likely most of you did not use line weight in AutoCAD because of these issues.

Try out the new MicroStation method. You will probably like it. Remember, you can turn off line weight displays using View Attributes, and the drawing view will look even closer to what you are used to in AutoCAD.

Level Symbology

One of the best level features is the ability to define level symbology on a per-level or per-file basis. It is recommended that you investigate level symbology and work with it until you fully understand its capabilities. This is such a useful feature that the time will be well spent.

Level symbology is defined in the Level Manager dialog by selecting the Symbology button and activating the Overrides setting. Each level has in its definition the ByLevel definitions for color, style, and weight and the Override definitions for color, style, and weight. These settings are saved with the design file for use throughout the life cycle of the file. Default symbology settings can be saved in seed files or DGNLIB files for individual standards, project standards, and corporate standards. Check out the exercise at the end of the chapter to fully understand this feature.

AutoCAD Command Comparison

AutoCAD	MicroStation
VISRETAIN	Use level symbology overrides located in the Level Manager dialog.
Layer States	Use level symbology overrides located in the Level Manager dialog along with saved views to memorize layer states.

Purge Levels versus Compress

The Purge Levels command is *not* what you think! It is not equivalent to the Purge Layers command found in AutoCAD that cleans up unused layer data in a DWG file.

Purge Levels will delete levels currently being used. This is a significant difference. The tool you are really looking for in MicroStation is Compress Design. The Compress design tool allows you to clean up unused data in a DGN file. Use the

Compress dialog options to define what data you want to clean up in the DGN file. All AutoCAD users need to be careful of this one or data will disappear inadvertently.

AutoCAD Command Comparison

AutoCAD	MicroStation
PURGE **File > Drawing Utilities > Purge**	Use the **File > Compress > Options** and **File > Compress > Design** tools to clean up unused data in a DGN file.

In Exercise 6-1, following, you have the opportunity to practice manipulating levels in a single view.

EXERCISE 6-1: MANIPULATING LEVELS IN A SINGLE VIEW

In this exercise you will learn to create levels with or without ByLevel definitions, as well as how to manipulate which levels are displayed in a view.

1   Open the design file *LEVELS_1.DGN*.

2   Open the Level Manager dialog found on the Primary Tools toolbar.

3   Click on the New Level button to create a new level.

Key in the level name *Buildings* and set the following attributes to this level.

Color: 1   (blue)
Style: 2   (dashed)
Weight: 3

To change the current level attributes, select the current level color, style, and weight. This will activate the applicable pop-up dialog for making the changes.

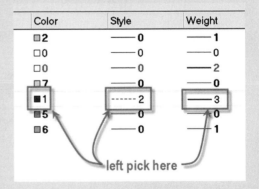

4   Double click on the *Buildings* level name to make it the active level.

It should turn an aqua color.

5   Close the Level Manager dialog.

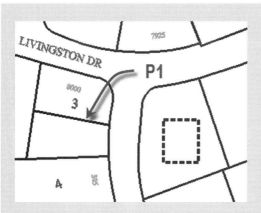

6 Draw a building outline on one of the lots using this new level. The building line attributes should match the new level settings: blue, dashed, and a heavy line weight.

7 Change the active level using the Attributes toolbar.

Select the Level pull-down menu and select the level *LOT_LINES* to add an additional lot line to the drawing.

Add the lot line identified as P1 using the keypoint snap.

8 Turn off the *Buildings* level using the Attributes toolbar.

9 Delete the new lot line drawn in step 7.

You will find that you cannot. Why not?

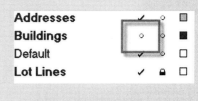

10 Check the level status of *LOT LINES* in the Attributes toolbar. Do you see it?

Yes, the level is locked and therefore you cannot modify anything on that level. Note the message in the status bar message center.

11 Unlock the *LOT LINES* level and try to delete the line again. You should be able to modify elements on that level.

In Exercise 6-2, following, you have the opportunity to practice manipulating levels in multiple views.

EXERCISE 6-2: MANIPULATING LEVELS IN MULTIPLE VIEWS

In this exercise you will learn to turn levels on and off using multiple views and files.

1 Open the design file *LEVELS_2.DGN*.

2 Turn on view windows 1 and 2.

3 Tile both view windows using the pull-down menu **Window > Tile**.

Open the Level Display dialog for this next step.

4   Use the shortcut key combination Ctrl + E and place this dialog between the two open view windows so that you can see the drawing in both windows.

Turn on the *Buildings* level in view window 2 only.

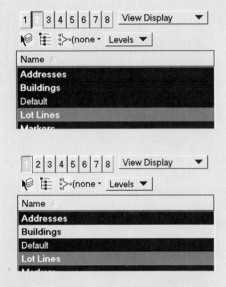

5   Verify that only view number 2 activated in the View Index buttons, as shown in top left figure.

Then select the *Buildings* level to turn it on. Remember, BLACK represents ON.

6   Turn off view 2 in the View Index buttons and turn on view 1, as shown in bottom left figure.

Note that the *Buildings* level is still off for this view.

Let's turn off the *Plot Numbers* level in all view windows.

7   Activate all view numbers in the View Index buttons (as shown at right).

8   Turn off the level *Plot_Numbers*.

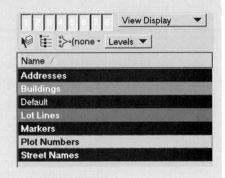

In Exercise 6-3, following, you have the opportunity to practice manipulating levels in multiple files.

EXERCISE 6-3: MANIPULATING LEVELS IN MULTIPLE FILES

In this exercise you will learn to turn levels on and off using multiple views and files.

1   Open the design file *LEVELS_3.DGN*.

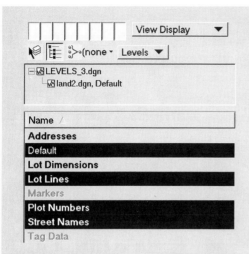

Now we want to turn on the level *Plot Numbers* in just the active file, not the reference file.

2 Verify that the File Target Tree option is activated so that you see the list of all available files.

3 Select the reference file *LAND2.DGN*.

4 The level *PLOT NUMBERS* is dark gray, representing that its display varies in the activated views. That is okay. We will just toggle it off and then back on in all view windows.

5 Select the active file *LEVELS_3.DGN* and turn on the *Plot_Numbers* level in all views.

In Exercise 6-4, following, you have the opportunity to practice using level symbology.

EXERCISE 6-4: USING LEVEL SYMBOLOGY

In this exercise you will learn to set up and use level symbology.

1 Open the design file *LEVELS_4.DGN*.

Next, we want to define and use level symbology in the active file only.

2 Open the Level Manager dialog via the Primary Tools toolbar.

3 Select the Symbology setting for Overrides and review the settings on all levels. Many times, the defaults are all set to "0" for color, style, and weight.

Let's see what this changes in our view windows.

4 Open the View Attributes dialog using the shortcut key combination Ctrl + B.

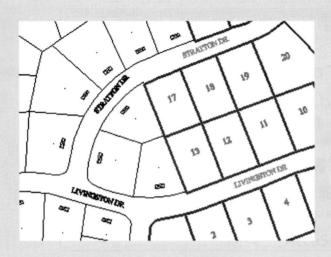

5 Turn on the Level Symbology setting for view window 1 only.

Verify that the View Index has view 1 activated, and then click on the Apply button to apply this change to the view.

The symbology changes are applied to view 1 for the active file only. Why? If you said because of the files selected in the Level Manager window you were correct. Go back to the Level Manager and check the Override settings for the reference file *LAND2.DGN*. Make sense now? Good! So, how would you apply the same override settings from the active file to the reference files?

6 Select the reference file in the file list and set the override colors, styles, and weights to 0.

You might need to use the Update View command to see this change.

For the next few steps, make the following symbology changes. Try it without instructions, but if you need help refer to steps 7 through 12.

LOT LINES (all files):
> *Color:*   3   (red)
> *Style:*   2   (dashed)

LOT LINES (active file only):
> *Weight*:   4

7 Open the Level Manager and select all files in the file list.

8 Verify that you are changing the Override symbology.

9 Select the level *LOT LINES* and change the attributes as requested.

10 Select the active file only (*LEVELS_4.DGN*).

11 Select the level *STREET NAMES* and change the attributes as requested.

12 Close the Level Manager dialog.

Did you get the result shown at left?

You can turn off level symbology without removing the settings defined in overrides. Open the View Attributes dialog and toggle off level symbology. Be sure to click on the Apply button.

Toggle level symbology back on to use the overrides again. Use the All button to apply changes to both view windows. Can it get any simpler than that?

USING REFERENCES

The use of reference files should not be new to the majority of AutoCAD users, so in this section we will discuss the differences between MicroStation's reference files and AutoCAD's Xrefs.

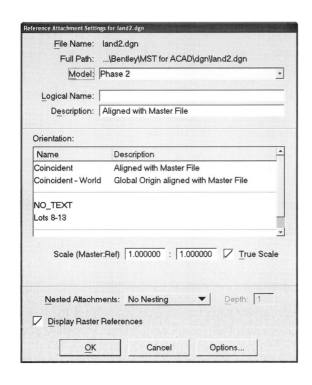

Attach a Reference

There are several decisions to be made when you attach a reference file.

1   Which model in the design file should you attach?

2   Do you need a logical name?

3   What type of attachment should you use? Coincident? Coincident World? Saved View?

4   Should you use True Scale or not?

5   What type of nesting is appropriate?

These questions are probably not something you can answer immediately. Keep reading.

MODEL

The Model option allows you to select any model within the referenced file as your attachment. Remember, the DGN file can contain multiple models (modelspaces) within a single file.

LOGICAL NAME/DESCRIPTION

This information is optional and not required for attaching a reference file, but it can be very beneficial down the road when you want to manipulate the reference attachment. Using a logical name simplifies future editing and assists other users in understanding what each reference is being used for. If the logical name is not enough, specify more information in the description. Keep logical names short but descriptive for greater efficiency. If you attach the same file more than once, a unique logical name is automatically assigned to the second attachment and all subsequent attachments.

ATTACHMENT MODE

When attaching a reference file, there are several methods available that control the geographic relationship between the active file and the attached reference file. They are coincident, coincident to world, and saved views.

Coincident: Attaches the reference file to the active file by matching up the design plane coordinates, but not the global origin 0,0,0.

Coincident World: Attaches the reference file to the active file by matching up the design plane coordinates and the global origin 0,0,0.

Saved View: Attaches the reference file using a saved view from the reference file. This method is interactive and the coordinate systems have no impact on the location. This is a good way of controlling which levels and geographic areas are displayed in the attachment.

TIP: *You can use the Saved Views option to pre-clip your reference file and pre-set your level displays prior to attaching the file as a reference file.*

SCALE (MASTER:REF)

The Scale option Master:Ref allows you to scale the reference file during the attachment. This defines the "ratio" of active (master) file working units to reference file working units.

TRUE SCALE

The True Scale option allows you to match the working units from the reference file with your active file automatically based on the scale specified. This means you could attach a metric reference file (mm) to an Imperial active file (inches) and the reference file would adjust by 25.4 as needed so that the units of 1 are now equal. The good news is that you do not have to be a mathematician and figure all of this working unit stuff out anymore. Just let True Scale do its job. It works, and if you don't believe me you can get out the calculator and figure it out manually. Just be thankful you don't have to worry about the units in your file, or the units in that file you got from someone else who used some other units and some other software at some other company.

NEST DEPTH

Depth: `1`

The Nest Depth setting allows you to control what reference hierarchy is inherited automatically, and how deep the files are in the tree when inherited.

0: Do not inherit any reference files (similar to Overlay in AutoCAD)

1: Inherit one tree level of reference files

2: Inherit two tree levels of reference files

99: Inherit 99 tree levels of references files (similar to Attachment in AutoCAD)

NOTE: *Do* not *attempt to inherit 99 references. It is used here as an example of what* not *to do.*

One of the most powerful aspects of this concept is the ability to inherit enough information to be beneficial but not so much as to make it counterproductive.

▶ No Nesting
Live Nesting
Copy Attachments

NESTED ATTACHMENTS

Nested Attachment functionality provides various methods for controlling how inherited reference files are attached.

No Nesting: No attachments in the reference file will be inherited with this attachment.

Live Nesting: Reference attachments found in the reference file will be inherited based on the nest depth factor. The parent/child relationships are maintained.

Copy Attachments: Reference attachments found in the reference file will be inherited based on the nest depth factor. The parent/child relationships are not maintained.

SCALE LINESTYLES

The Scale LineStyles option provides the ability to scale reference files' line styles independently of the active file. This is especially important when referencing other file formats such as DWG that use line styles differently than the DGN file format.

ON: Custom line style components are scaled by the Scale (Master:Ref) factors.

OFF: Custom line style components are not scaled.

TIP: *Another method of attaching reference files is to drag and drop them from Windows Explorer into the Reference File dialog. The Attachment dialog will automatically appear if you drag in just one file. If you drag in multiple files, they will all be attached coincidentally and you must modify them after the attachment.*

MISSING ATTACHMENTS

Red reference file attachments indicate a "missing" reference file. This can be caused by various situations.

❑ File path is no longer valid

❑ File name is no longer valid

❑ File no longer exists (has been moved or deleted)

Reference Options

There are additional options available when attaching a reference file.

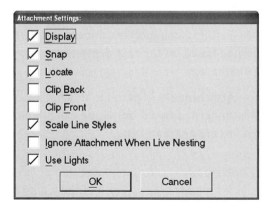

CLIP BACK

The Clip Back option sets the back clipping plane for 3D models.

CLIP FRONT

The Clip Front option sets the front clipping plane for 3D models.

IGNORE ATTACHMENT WHEN LIVE NESTING

The Ignore Attachment when Live Nesting setting allows you to control whether this file will be handled as a nested reference file when using the Live Nesting option. This is just another way of controlling individual files within the live nesting environment.

USE LIGHTS

Use the Use Lights option to control whether lights defined in the reference file models are used during rendering view operations.

Reference Dialog

The sections that follow describe options available in the Reference dialog.

DISPLAY

The Display option controls whether or not reference files are visible. It is useful for turning off the display of a reference temporarily, which is much more useful than "detaching" the reference file and later "reattaching" the file when the display is needed again. All reference file settings are retained while the display is off.

SNAP

The Snap option allows you to snap to elements in the reference file while placing or manipulating the active file elements. Deactivate this setting for reference files you need to avoid snapping to.

LOCATE

The Locate option allows you to copy reference file elements into the active file. It does not detach the original reference file attachment. You must detach the reference file separately if it is no longer needed. You can copy individual elements or multiple elements by using Fence or Selection Set tools.

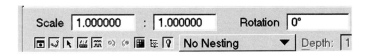

Reference Settings

| Attachment... |
| Presentation... |
| Update Sequence... |
| Adjust Colors... |
| Hilite   ▸ |
| Auto Arrange Icons |
| Level Manager |
| Level Display |

The sections that follow describe options available from the Settings menu in the Reference dialog.

ATTACHMENT

The Attachment settings allow you to modify the attachment settings after the initial attachment is made. You can modify the file path, logical name, and description.

PRESENTATION

The Presentation settings allows you to set a reference file's view rendering mode and other parameters, such as wireframe, wire mesh, hidden line, filled hidden line, and different types of shading.

UPDATE SEQUENCE

The Update Sequence settings are useful when using opaque fills or heavy line weights. It controls the visual draw order of the reference file to the active file. Bringing files with opaque fills to the top of the update sequence will place them as the "first" file to be updated, and they will appear on the bottom of the other files in the list.

ADJUST COLORS

The Adjust Colors settings allow you to adjust the hue and saturation value of the reference file colors. You can, for example, assign one fixed color such as grayscale to one reference file and a different fixed color (such as red) to all other files. This is somewhat similar to level symbology, except that it is significantly easier to define. You can print these adjusted colors, hues, and saturation values if needed.

HILITE

Use the Hilite settings to define how you want the reference files selected in the References dialog box to be displayed in the view window. You can highlight the entire file, just the boundary or clipping edge, or both. This is extremely helpful when using the References dialog and you are unsure which reference file you are looking at in the view window.

AUTO-ARRANGE ICONS

The Auto-Arrange Icons setting will change how the Reference tools move tools around when the dialog is resized.

LEVEL MANAGER

The Level Manager settings give you easy access to the Level Manager dialog and its reference file capabilities.

LEVEL DISPLAY

The Level Display settings give you easy access to the Level Display dialog and its reference file capabilities

INFORMATION PANEL

The Information panel is new in V8 and provides quicker access to common tools and settings on the selected reference files. For example, you can quickly change the attachment type using this panel. Note that this panel may disappear completely if the References dialog is resized too small.

Self-references

One of the biggest differences in reference files between MicroStation and AutoCAD is in what files you can attach. In MicroStation, you can attach a file to itself, commonly referred to as a self-reference attachment. Why would you want to do this, you ask? One example would be to clip different areas of a design at different scales. Say you have a layout and you know it is going to continue to change throughout the project. You want to have an expanded view of an area in your active design file. Using self-references allows you to attach a file to itself, clip it and scale it up, and move it to the side. All of the information in the clipped reference file is actually elements in your active design file.

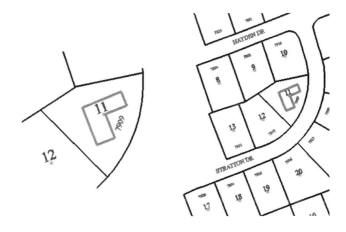

In Exercise 6-5, following, you can practice using references.

EXERCISE 6-5: USING REFERENCES

In this exercise you will learn to attach both DGN and DWG reference files, and apply level symbology and level commands to both. You will also learn how to clip a reference so that you can use just a portion of the file.

1   Open the design file *REFERENCES_1.DGN*.

2   Open the Reference dialog via one of the following methods.

Select the References button on the Primary Tools toolbar.

Go to the pull-down menu **File** > **References**.

3   Attach a reference file via one of the following methods.

Select the Attach Reference button.

Go to the pull-down menu **Tools** > **Attach** in the References dialog.

Right-click in the References dialog and select the Attach tool from the pop-up menu.

4   Select the file *REFERENCES_2.DGN* and click on Open to attach the file.

5   Establish the following settings for this reference file attachment.

Model:	Survey
Logical Name:	Surv
Description:	Final Survey Data
Orientation:	Coincident – World
Scale:	1:1
True Scale:	ON
Nested Attachments:	No nesting
Display Raster References:	OFF

Click on the OK button to complete the attachment.

HINT: *You can modify most of these settings using the Information panel at the bottom of the References dialog.*

Now, let's attach a DWG file as a reference file.

6   Access the Attach Reference tool and select the file *REFERENCES_3.DWG*.

When a DWG file has been selected as the reference attachment, the DWG Options button is available. The default settings under DWG Options should be adequate for most attachments, but if you have a problem revisit these settings.

7 Click on the Open button to complete the attachment.

8 A dialog asking for Unit Conversion information will display (similar to when you open a DWG drawing file).

What is the base unit definition of the DWG file? In this case the answer is feet.

9 Establish the following settings for this reference file attachment.

Model:	Model
Logical Name:	DWG contours
Description:	DWG from Company ABC
Orientation:	Coincident – World
Scale:	1:1
True Scale:	ON
Nested Attachments:	No nesting
Display Raster References:	OFF

Click on the OK button to complete the attachment.

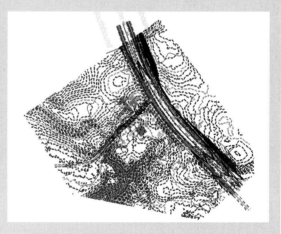

10 The final result should look as shown in the figure at right.

11 Practice turning levels on and off using the commands Off By Element, All Off, and All On.

HINT: *Look in the Level Display dialog. If that isn't enough of a hint, try the right-click on top of the level list.*

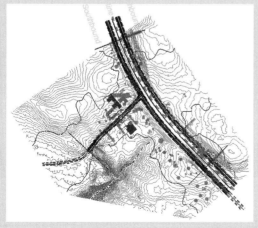

12 Practice applying level symbology to the DWG file contours and change the Symbology Override settings to the following.

Color: 9 (gray)
Style: 0 (continuous)
Weight: 0

HINT: *If you can't remember how to do this, see Exercise 6-4.*

Next, you will learn how to clip and edit a reference file in place.

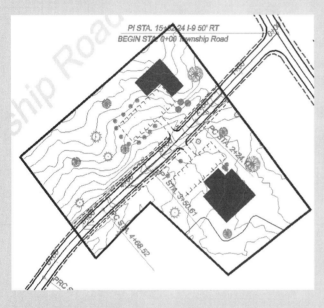

13 Turn on the level *Clip Fence Shape.*

14 Place a fence using the element Fence Type and the Clip Fence Mode tool settings.

Select the new element boundary for the fence shape.

15 Open the References dialog and highlight both reference files.

16 Right-click to access the Clip Boundary command, and use the Fence Method tool setting.

Issue a data point in view 1 to complete the Clip command.

17 Turn off the level *Clip Fence Shape.*

Now you can see the clipped reference files. Note that the active file's elements were not clipped.

18 To delete a clip boundary, use the Delete Clip command found in the References dialog using **Tools > Delete Clip**. This command is applied to all selected references in the dialog file list.

TIP: *You can clip the active file using the Clip Volume view command in conjunction with the Clip Volume view attribute.*

7: Working with Annotation

CHAPTER OBJECTIVES

- ❏ Learn to place annotation using text and dimensions
- ❏ Learn to use tag data for intelligent annotation
- ❏ Learn to use styles to automate standard annotation settings
- ❏ Learn to use annotation scale to automate text sizes

Placing annotation in an engineering drawing is always a combination of technical know-how and skillful planning. Finding room for all of the necessary text and dimensions is always a challenge, and predicting the size of these annotations is a guessing game at best. MicroStation provides several tools that assist with this process, some of which are highly automated and others that are electronic versions of the old hand drafting concepts.

TYPES OF TEXT ELEMENTS

There are several types of text in MicroStation, and understanding their differences is important. You can place text elements, text nodes, enter data fields, notes, dimensions, tags, and flags.

Text Elements

The *text* element is a single line of text that can be placed using the Place Text command. All text elements can contain up to 65,535 characters, if you can find room for text strings that long.

Text Nodes

The *text node* element is a multi-line paragraph of text that is also placed using the Place Text tool. Text nodes are generated automatically when the Enter key is used to indicate a second line of text while typing in your text data.

Text nodes can display a text node view attribute, which is small crosshair and a number, if enabled in the view window attributes. These text node numbers will plot if inadvertently left on.

> **TIP 1:** *Use view attributes to toggle the display of text nodes in the view window.*

> **TIP 2:** *Use plot configuration files to automatically disable the ability to plot text nodes.*

Enter Data Fields

The *enter data field* element is unique to MicroStation in its ability to provide easy editing features, and powerful text manipulation tools. An enter data field is recognizable by its distinctive "underscore" appearance, and the rectangular box that appears in association with text editing tools. This element type allows you to place a piece of text that defines all properties of the text feature (font, size, level, color, weight, and line style) but without the actual text characters. It is ideal for data that is annotated on the drawing, but has yet to be determined through design, such as equipment numbers, room numbers, title block information, and generic text in standard symbols.

Notes

The *place note* element is actually a dimension element similar to the AutoCAD leader. This element contains text and other graphics features such as lines, arcs, and terminators. The text portion of this element is "associated" with the graphical components, providing easy editing later. The appearance of this element is controlled by a dimension style or independent dimension settings.

Dimensions

The *dimension* element provides graphic-based annotations whose primary purpose is to display "real-world" sizes and their related text.

Tags

The *tag* element is basically a piece of text that has some additional intelligence associated with it. This element can be associated with individual graphics or symbols and can be globally edited or extracted for use outside the design file. This is not a widely used annotation feature, but it is the closest relative to AutoCAD's attribute text found in blocks.

Flags

The *flag* element is a bitmap icon image that acts like an electronic sticky note. You can place a flag in your drawing and key in associated notes, which provides a valuable method for distributing information between users of the design file.

Placing Text

PLACE TEXT

Use the Place Text tool to place single-line and multi-line text elements. The Tool Settings dialog provides several options for controlling the appearance and location of a text element.

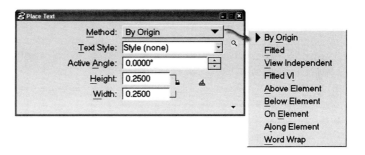

1   Select the Place Text tool.

2   Modify the necessary tool settings.

3   Key in the text characters.

Methods. Use the text method to determine how the text is placed.

By Origin: Places text by a justification point.

Fitted: Places text between two data points.

View Independent: Places text independent of view rotation.

Fitted VI: Places fitted text independent of view rotation.

Above element: Places text above an element.

Below element: Places text below an element.

On Element: Places text on an element, breaking the element automatically.

Along element: Places text along a curved element, either above or below it.

Word Wrap. Automatically wraps text based on a rectangular area specified by the user. This method is only available if you are using the word processor text editor.

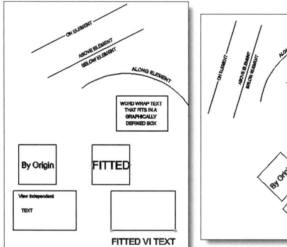 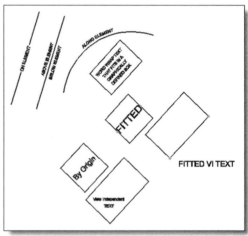

Text Style. Use the Text Style tool setting to select the style of text you want to place. Refer to the "Text Styles" section later in this chapter for additional information on text styles.

Active Angle. The Active Angle setting defines the angle to be used when placing elements.

Height. The Height setting defines the height of the text characters. Use the Lock icon to lock the height and width together for proportional text characters.

Height:	0.2500
Width:	0.1250

Width. The Width setting defines the width of the text characters. Use the Lock icon to lock the height and width together for proportional text characters.

AutoCAD Command Comparison

AutoCAD	MicroStation
Mouse pick: Keyboard: *MText*	Use the Place Text tool to place multi-line strings of text.
Height	Use the Height and Width settings.
Justify	Use the Justification setting.
Line spacing	Use the Linespacing setting.
Rotation	Use the Active Angle setting.
Style	Use the Text Style setting.
Width	The width of the paragraph is controlled by the Line Length setting, which is not available in the Tool Settings dialog. Use the key-in *LL=40* to define a line length of 40 characters. Use the Word Wrap option to graphically define the paragraph width.

EDIT TEXT

Use the Edit Text tool to edit any existing text element. The various tool settings allow you to modify text attributes at the same time you are editing the text content.

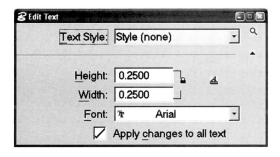

1 Select the Edit Text tool.

2 Select an existing text element to edit.

3 Define the tool settings as needed.

You can edit one text element at a time by selecting it individually, or several text elements defined by a selection set or fence.

TIP 1: *Enter data fields will display with <<TEXT>> in the editor window. If you remove the << or >> characters you remove the enter data field capabilities and convert the text to a normal text element.*

TIP 2: *MicroStation provides for double-click editing through the use of the Selection tool. This feature is available for text and dimension text only.*

AutoCAD Command Comparison

AutoCAD	MicroStation
Mouse pick: Keyboard: DDEDIT Edit Text	Use the Edit Text tool to edit existing text elements.
Undo	Use the Undo button or Ctrl + Z to execute the undo in the middle of the command.

SPELL CHECKER

Use the Spell Checker tool to check your spelling throughout the entire drawing.

1   Select the Spell Checker tool.

2   Identify the text elements to be checked.

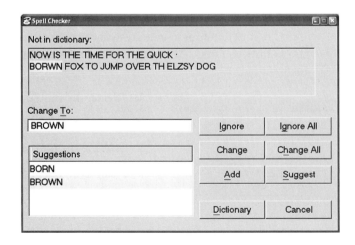

You can spell check one text element at a time by selecting it individually, or several text elements defined by a selection set or fence.

AutoCAD Command Comparison

AutoCAD	MicroStation
Keyboard: *SPell*	Use the Spell Checker tool to verify correct spelling in existing text elements.

DISPLAY TEXT ATTRIBUTES

Use the Display Text Attributes tool to display an existing element's text attributes in the status bar at the bottom of the application.

TH = text height

TW = text width

LV = level

FT = font

1 Select the Display Text Attributes tool.

2 Identify an existing text element.

Sorry, those are the only text attributes that display with this function.

AutoCAD Command Comparison

AutoCAD		MicroStation
Mouse pick:  Properties	Keyboard: *PRoperties*	Use the Display Text Attributes tool to display just the text settings of an existing element. Use the Element Information tool to display all settings of an existing element.

Display Text Attributes > TH=0.2500, TW=0.2500, LV=64, FT=Arial

MATCH TEXT

Use the Match Text tool to match all text attributes from an existing text element. This tool will modify all active text attributes so that all text placed from this point forward will look identical to the existing text element you selected.

1   Select the Match Text tool.

2   Identify an existing text element.

3   Issue a data point to accept.

Text attributes are as follows.

❑ Text style	❑ Font
❑ Text height	❑ Text width
❑ Line spacing	❑ Line space type
❑ Slant	❑ Line length
❑ Underline	❑ Vertical text
❑ View dependency	❑ Intercharacter spacing
❑ Justification	

The Match Text tool will not match any other attributes (such as level, color, line weight, and so on) associated with the text element.

> **TIP:** *The concepts used in all MicroStation match commands are completely opposite those found in AutoCAD match properties functionality. In AutoCAD, you place the text first and then match it to an existing object to inherit property settings. In MicroStation, you should match the existing text element first, and then place a new text element using the active text attributes that now match the existing text element you selected.*

AutoCAD Command Comparison

AutoCAD		MicroStation
Mouse pick	Keyboard: *MAtchprop*	Use the Match Text Attributes tool to set the active text attributes to those of an existing element.
Match Properties		Use the SmartMatch tool to set *all* active element attributes to those of an existing element.

CHANGE TEXT ATTRIBUTES

Use the Change Text Attributes tool to modify the text attributes for any existing text elements. This tool will modify "text" attributes only.

1 Select the Change Text Attributes tool.

2 Define the tool settings changes.

3 Identify an existing element.

You can change one text element at a time by selecting it individually, or several text elements using a selection set or fence.

AutoCAD Command Comparison

AutoCAD	MicroStation
Mouse pick: Keyboard: ![Properties icon] *PRoperties* Properties	Use the Change Text Attributes tool to modify just the text properties of an existing element.

PLACE TEXT NODE

The Place Text Node tool is rarely used today, but is a legacy command from previous versions. It still works, however, and you can place a text "marker" in your drawing and then populate it with characters later. Text nodes consist of a crosshair and a text node number. They will plot if the *view window* attribute is active. Many users prefer this method to *enter data fields* because the justification is easier to control with *text node* elements.

You can place view-independent text nodes as follows.

1 Select the Place Text Node tool.

2 Identify the text node location in the drawing.

AutoCAD Command Comparison

There is no equivalent command in AutoCAD for Place Text Node.

COPY/INCREMENT TEXT

The Copy/Increment Text tool allows you to copy an existing text element and increment its value by any increment value. You can key in positive and negative increment values.

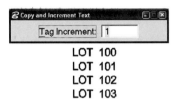
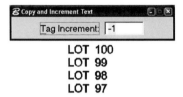
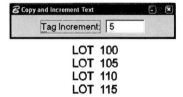

1 Select the Copy/Increment Text tool.

2 Define the increment value in the Tool Settings dialog.

3 Select an existing text element as the base value.

4 Identify a new location to copy the new incremented text to.

AutoCAD Command Comparison

AutoCAD *Express Tool*	MicroStation
Keyboard: *TCOUNT*	Use the Copy/Increment Text tool to automatically place text with incrementing values.

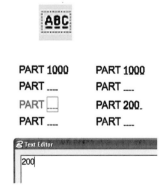

FILL IN ENTER DATA FIELD

Use the Fill in Enter Data Field tool to fill in empty data fields or to modify already filled enter data fields.

1 Select the Fill in Enter Data Field tool.

2 Select a data field.

3 Key in the data field value

You must select the data fields manually using this tool.

AutoCAD Command Comparison

There is no exact equivalent for enter data fields. However, when enter data fields are used in cells, the equivalent command would be block attributes in AutoCAD. Refer to the Place Tag tool in MicroStation for the exact equivalents of block attributes.

AUTO-FILL ENTER DATA FIELD

Use the Auto-Fill Enter Data Field tool to fill in empty data fields automatically.

1 Select the Auto-Fill Enter Data Field tool.

2 Select a view window to begin searching for empty data fields. Continue to issue data points in the view window until the correct data field highlights.

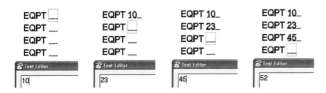

3 Key in the data field value, and press the Enter key to fill in the value and to move on to the next empty data field.

4 Continue to key in values (using the Enter key to move to the next data field).

AutoCAD Command Comparison

There is no equivalent command in AutoCAD for Auto-Fill Enter Data Field.

COPY ENTER DATA FIELD

Use the Copy Enter Data Field tool to copy enter data field values from one data field to another. This can be very useful when you have repetitive part numbers throughout a drawing layout.

1 Select the Copy Enter Data Field tool.

2 Select the data field you want to copy from.

3 Select the data field you want to copy to.

PART 1000	EQPT __
PART __	EQPT __
PART __	EQPT __
PART __	EQPT __

PART 1000	EQPT 1000
PART __	EQPT __
PART __	EQPT 1000
PART __	EQPT __

AutoCAD Command Comparison

There is no exact equivalent for enter data fields. When enter data fields are used in cells, the equivalent command would be block attributes in AutoCAD. Refer to the Place Tag tool in MicroStation for the exact equivalents of block attributes.

COPY/INCREMENT ENTER DATA FIELD

Use the Copy/Increment Enter Data Field tool to copy a data field value to another value and increment it by a defined value at the same time. You can key in positive and negative increment values.

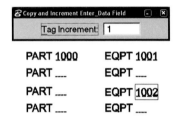

1   Select the Copy/Increment Enter Data Field tool.

2   Define the Increment Value tool setting.

3   Select the data field you want to copy and increment from.

4   Select the data field you want to increment.

AutoCAD Command Comparison

There is no exact command in AutoCAD for Copy/Increment Enter Data Field.

XYZ ANNOTATION

If you work in the civil or survey disciplines, you need to know about the XYZ Text toolbar. The coordinate tools of this toolbar provide methods of importing and exporting coordinates in a design file, and for labeling XYZ coordinate locations.

LABEL COORDINATES

Use the Label Coordinates tool to label specific XYZ coordinates in your design file.

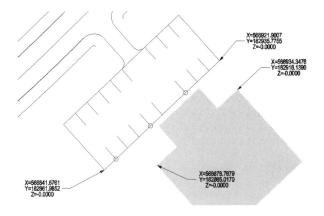

1 Select the Label Coordinates tool.

2 Issue a data point to label the coordinate.

3 Use a Tentative point or AccuSnap for exact precision when identifying these coordinate points.

TIP: *Edit the coordinate text and use the Place Note tool to add the leader to the coordinate text.*

AutoCAD Command Comparison

There is no equivalent command in AutoCAD for Label Coordinates.

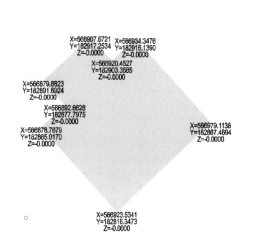

LABEL ELEMENT

Use the Label Element tool to label the coordinates of all vertices on an existing element.

1 Select the Label Element tool.

2 Click on the Single button and identify the element whose coordinates you need labeled.

3 Issue a data point to accept the element and place the coordinates.

Note that you can label multiple elements using the Fence and All tool setting options.

AUTOCAD TIP: *AutoCAD users may prefer to use the Label command to access information, similar to using the List command found in AutoCAD. You do not have to actually place the label to see the values for length and angle.*

AutoCAD Command Comparison

There is no exact command in AutoCAD for Label Element.

IMPORT AND EXPORT COORDINATES

Use the Import and Export Coordinates tool to import and export coordinates to a comma- or space-delimited text file. These types of files can include a single element, multiple elements, or all elements in the design file. The following figures show examples of the building coordinates exported to a comma-delimited file. You can also import these coordinates into another design file.

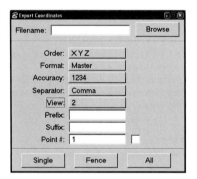
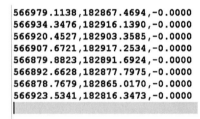

The following figure shows the building survey points after the import. These points can be imported as point elements, text elements, or cells.

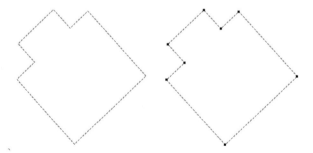

AutoCAD Command Comparison

There is no equivalent command in AutoCAD for Import and Export Coordinates.

DRAFTING TOOLS

There are several annotation drafting tools for automating common bubbles, titles, and callout text. The appearance of these symbols is controlled by the Parameters tool located on the Drafting Tools toolbar. These tools create the following standard non-associated drafting symbols.

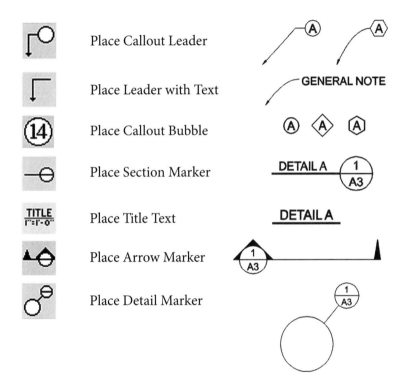

Place Callout Leader

Place Leader with Text

Place Callout Bubble

Place Section Marker

Place Title Text

Place Arrow Marker

Place Detail Marker

DRAFTING TOOL PROPERTIES

The dialog at right displays the various settings available for placing all Drafting Tool annotation elements.

AutoCAD Command Comparison

There are no equivalent commands for drafting annotation provided in AutoCAD.

TEXT STYLES

Text Styles are a new feature in V8, and thus much of the text found in legacy DGN files will not have text styles defined, and the majority of the existing text probably will not be taking advantage of this feature. Text styles store all settings that control what a piece of annotation looks like. They provide independent and global editing capabilities, making last-minute changes easier to accomplish.

Another added benefit to using Text Styles is the "linking" functionality associated with using styles from a style library. Style libraries are files (typically with the *.DGNLIB* extension) that contain standard styles for text, dimensions, levels, and multi-lines. These style libraries can also be defined in a spreadsheet file. Style libraries allow you to control CAD standards used in design files.

A text style library automates text settings, so that you are less likely to use a nonstandard font or to guess at the appropriate text size. Styles can be imported or attached to encourage following a corporate standard. You are strongly urged to give styles of any type due consideration. They are real time savers. Text styles are very easy to set up and to use, so let's take a look at some examples.

Non–Scale-based Text Styles

This type of text style defines the basic settings for the text elements but does not define text sizes for specific output scales. A style defined in this way would be used as a generic base for all text elements. However, you would be required to size the text manually or with annotation scale during placement.

Style Name	GENERAL NOTES
Parent Style Name	- None -
Font	WORKING

- GENERAL NOTES
- PRESENTATION
- TITLE TEXT

GENERAL NOTES

	Style Name	PRESENTATION
⅃ GENERAL NOTES	Parent Style Name	- None -
⅃ PRESENTATION	Font	⊤ Arial
⅃ TITLE TEXT		

PRESENTATION

	Style Name	TITLE TEXT
⅃ GENERAL NOTES	Parent Style Name	- None -
⅃ PRESENTATION	Font	⅃ complex
⅃ TITLE TEXT		

TITLE TEXT

TIP: *Settings in blue have not been saved back to the style and can be used as a temporary override. Save the style to make temporary overrides permanent, or reset the style to discard these temporary overrides.*

Scale-based Text Styles

You can create Text Styles using a parent/child relationship to automatically set the text heights for all standard text. This type of text style defines the basic settings for the text elements and the exact text sizes for specific output scales. A style defined in this way would be used as an automated method for placing text elements; and you would not be required to size the text during placement. All text sizes are defined in the actual text styles.

The parent/child relationship ensures that settings defined by the parent would be automatically inherited by the child (unless the specific child were to define an override to the parent, in which case the child setting would control the parent). These terms can be explained using a typical family environment. Parents are in control until the child is created; and then the child overrides the parent. Pretty close to the truth, wouldn't you agree?

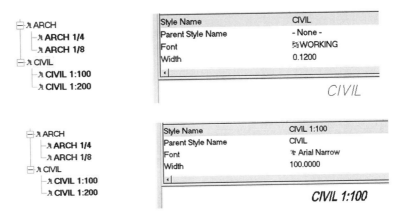

TIP: *Child settings in blue may have been saved. Be sure to check. When the blue highlight is on a child setting it usually represents the differences between the child and the parent, but blue also represents unsaved changes. An alternative color would be helpful, but for now they are both represented with a blue highlight color.*

AutoCAD Command Comparison

AutoCAD	MicroStation
Format > **Text Styles** Keyboard: *STYLE*	Use the **Element** > **Text Styles** selection to create and define text styles.

TYPES OF DIMENSIONS

There are several types of dimensions in MicroStation, and understanding their differences is important. You can place linear, angular, radial, and ordinate dimensions (along with their many tool settings) to label the size of just about anything in your drawing.

Dimension Tools Main

This tool frame contains all individual tools for dimensioning in MicroStation. It can be docked below the Main tool frame if you use these tools frequently.

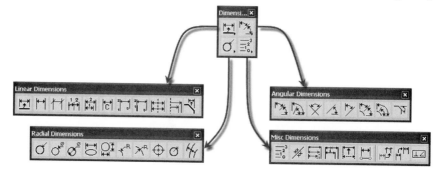

DIMENSION ELEMENT

The Dimension Element tool is a generic dimension tool that can place a dimension based on the type of element you select. This tool can only dimension a single element, and it cannot dimension between elements.

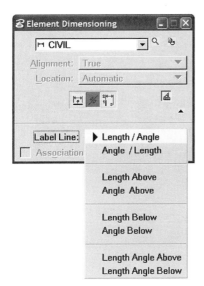

1 Select the Dimension Element tool.

2 Select the dimension style to be used.

3 Select the preferred Alignment method.

4 Select the dimension text Location.

5 Select the Dimension tool.

6 Activate annotation scale if needed.

Alignment. The Alignment setting determines which axis the dimension is aligned to.

> **View:** Aligns dimensions with the view independently of view rotation. Dimensions are aligned horizontal or vertical to the view.

> **Drawing:** Aligns dimensions with the unrotated view window and ignores any changes to the view rotation.

> **True:** Aligns dimensions with the element.

> **Arbitrary:** Aligns dimensions parallel to an element, but the extensions can be with any two points placed in the view window.

Location. The Location setting controls the location of the text in the dimension element.

> **Automatic:** Text is placed based on the justification setting

> **Semi-Auto:** Text is placed based on justification if the text fits appropriately. If not, you are prompted for text placement.

> **Manual:** Allows you to locate the text placement location by dragging the text manually.

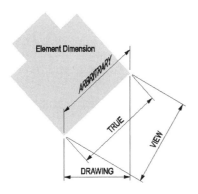

SITE PLAN

Dimension Element. The Dimension Element option dimensions an element differently depending on the type of element you select, such as a linear dimension for a line or a radial dimension for a curved element.

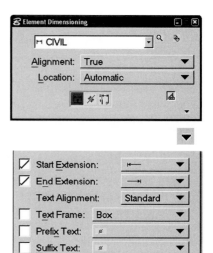

AutoCAD Command Comparison

AutoCAD	MicroStation
Mouse Pick: Keyboard: *DIMLINear* Linear	Use the Dimension Element tool to place a linear dimension on a linear element. Use the Diameter Parallel tool to place a linear dimension on a radial element.
Mtext	There is no equivalent option in MicroStation. You can edit the text in the dimension after placement.
Text	There is no equivalent option in MicroStation. You can edit the text in the dimension after placement.
Angle	There is no exact equivalent option in MicroStation. You can rotate the dimension text to vertical using the Text Alignment tool setting.
Horizontal	Drag the dimension using the mouse to control dimension axis.
Vertical	Drag the dimension using the mouse to control dimension axis.
Rotated	Use the Dimension Linear tool to place a rotated linear dimension. Control the angle using snap points on the dimensioned element.

Label Line. The Label Line option labels a linear or curved element with length and angle.

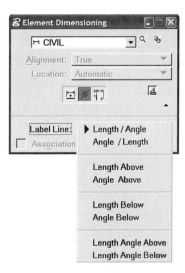

1 Select element to label with length and angle.

2 Issue a data point to accept label text placement or reset to abort.

AutoCAD Command Comparison

There is no equivalent command in AutoCAD for the Label Line tool.

 Dimension Size Perpendicular. The Dimension Size Perpendicular option labels the perpendicular distance between two elements.

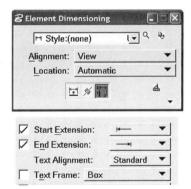

1 Select element to dimension.

2 Identify point or select element to dimension perpendicular to.

AutoCAD Command Comparison

There is no equivalent command in AutoCAD for the Dimension Size Perpendicular tool.

 ### DIMENSION LINEAR
A linear dimension is used to label the straight distance between or along elements.

1 Select the Dimension Linear tool.

2 Select the dimension Style to be used.

3 Select the alignment Method.

4 Select the dimension text Location.

5 If necessary, preset the dimension off-set distance from the element being dimensioned.

6 Select the dimension Type.

7 Activate the annotation scale if needed.

8 Define the remaining settings as needed.

9 Select the element to dimension.

10 Drag the dimension into position and issue a data point for the dimension location.

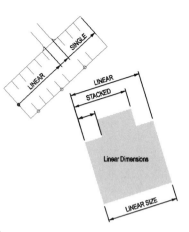

 Linear Size. The Linear Size tool will provide a running string of linear dimensions.

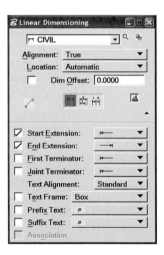

AutoCAD Command Comparison

AutoCAD	MicroStation
Mouse Pick: Keyboard: Linear	Use the Dimension Element tool to place a linear dimension on a linear element. Use the Diameter Parallel tool to place a linear dimension on a radial element.

Keyboard: *DIMLINear*

AutoCAD	MicroStation
Mtext	There is no equivalent option in MicroStation. You can edit the text in the dimension after placement.
Text	There is no equivalent option in MicroStation. You can edit the text in the dimension after placement.
Angle	There is no exact equivalent option in MicroStation. You can rotate the dimension text to vertical using the Text Alignment tool setting.
Horizontal	Drag the dimension using the mouse to control dimension axis.
Vertical	Drag the dimension using to control dimension axis.
Rotated	Use the Dimension Linear tool to place a rotated linear dimension. Control the angle using snap points on the dimensioned element.

Linear Stacked. The Linear Stacked tool provides a stacked series of dimensions. There is a significant difference in the resulting dimension element. The MicroStation stacked dimension is a single element, not individual dimension segments.

TIP: *You can use the Drop Element tool and the setting Dimensions > To Segments to explode the single dimension into individual segments.*

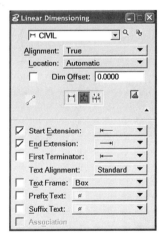

AutoCAD Command Comparison

AutoCAD	MicroStation
Mouse pick: Keyboard: *DIMBASEline* Baseline	Use the Linear Stacked tool to place baseline dimensions. You do not need a base dimension to continue from.

AutoCAD	MicroStation
Select	There is no need for this option because the stacked dimension in MicroStation does not require a base dimension.

Linear Single. The Linear Single tool provides a running string of dimensions that are all measured from a single point. The dimension text is a running total length.

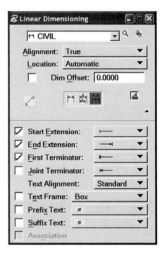

AutoCAD Command Comparison

AutoCAD		MicroStation
Mouse Pick: Continue	Keyboard: *DIMCONTinue*	Use the Linear Single tool to place continuous dimensions. You do not need a base dimension to continue from.
Select		There is no need for this option because the continuous dimension in MicroStation does not require a base dimension.

DIMENSION ANGULAR
An angular dimension is used to label the angle of (or between) elements.

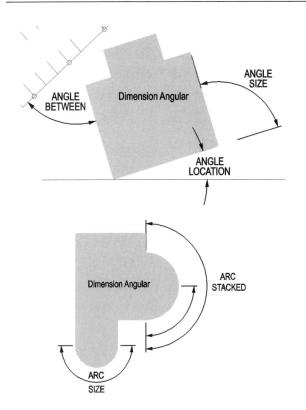

1 Select the Dimension Angular tool.

2 Select the Dimension Style to use.

3 Select the alignment method.

4 Select the dimension text Location.

5 Select the dimension Type.

6 Activate the annotation scale if needed.

7 Define the remaining settings as needed.

8 Select the element to dimension.

9 Drag the dimension into position and issue a data point for the dimension location.

Angle Size. The Angle Size setting will dimension the angle of an element calculated from the end point of the element.

Angle Location. The Angle Location setting will dimension the angle of an element calculated from the dimension origin.

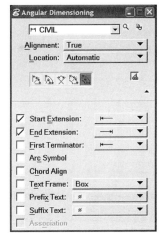

Angle Between. The Angle Between setting will dimension the angle between two elements.

Arc Size. The Arc Size setting will dimension the angle between end points on a curved element.

Arc Stacked. The Arc Stacked setting will dimension the angle between multiple points on an element.

AutoCAD Command Comparison

AutoCAD	MicroStation
Mouse pick: Keyboard: *DIMLINear* Angular	Use the Dimension Element tool to place a linear dimension on a linear element. Use the Diameter Parallel tool to place a linear dimension on a radial element.
Specify Vertex	Use the Dimension Angle Location tool to dimension an angle between elements using a specified vertex.
Mtext	There is no equivalent option in MicroStation. You can edit the text in the dimension after placement.
Text	There is no equivalent option in MicroStation. You can edit the text in the dimension after placement.
Angle	There is no exact equivalent option in MicroStation. You can rotate the dimension text to vertical using the Text Alignment tool setting.

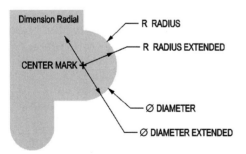

DIMENSION RADIAL

A radial dimension is used to label the size of circular and curved elements. You might consider making a specific dimension style for radial dimensions. This allows you to "tweak" the standard dimension style's appearance when used for radial dimensions.

1 Select the Dimension Radial tool.

2 Select the dimension Style to be used.

3 Select the preferred Mode.

4 Select the preferred Alignment.

5 Activate annotation scale if needed.

6 Select the circular element to dimension.

7 Drag the dimension into position and issue a data point for the dimension location.

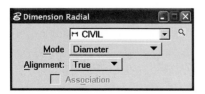

RADIUS

The Radius setting will dimension the radius of a circle or an arc element.

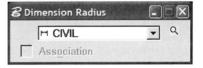

1 Select the Dimension Radius tool.

2 Select the dimension style to be used.

3 Select the circular element to be dimensioned.

4 Issue a data point for the dimension location.

EXTENDED RADIUS

The Extended Radius setting will dimension the radius of a circle or arc with an extended leader line.

1 Select the Dimension Extended Radius tool.

2 Select the dimension style to be used.

3 Select the circular element to be dimensioned.

4 Issue a data point for the dimension location.

AutoCAD Command Comparison

AutoCAD		MicroStation
Mouse pick: Radius	Keyboard: *DIMRADius*	Use the Dimension Element tool to place a quick radius dimension on a circle or arc. Use the tool setting Dimension Radius to control whether you get a radius or a diameter. Use the Radius or Extended Radius tool to place a radius on a circle or arc.
Mtext		There is no equivalent option in MicroStation. You can edit the text in the dimension after placement.
Text		There is no equivalent option in MicroStation. You can edit the text in the dimension after placement.
Angle		There is no exact equivalent option in MicroStation. You can rotate the dimension text to vertical using the Text Alignment tool setting.

AutoCAD Command Comparison

AutoCAD	MicroStation
Mouse pick: Keyboard: ![Jogged icon] *JOG* *DIMJOGGED* Jogged	Use the Dimension Element tool to place a quick radius dimension on a circle or arc. Use the Radius tool to place a radius dimension on a circle or an arc.
Center Location Override	This option is not needed because the MicroStation radius command does not place a center mark.

DIMENSION DIAMETER

The Dimension Diameter tool will dimension the diameter of a circle or arc element.

1 Select the Dimension Diameter tool.

2 Select the dimension style to be used.

3 Select the circular element to be dimensioned.

4 Issue a data point for the dimension location.

EXTENDED DIAMETER

The Extended Diameter tool will dimension the diameter of a circle or arc with an extended leader line.

1 Select the Dimension Extended Diameter tool.

2 Select the dimension style to be used.

3 Select the circular element to be dimensioned.

4 Issue a data point for the dimension location.

AutoCAD Command Comparison

AutoCAD	MicroStation
Mouse pick: Keyboard: ![Diameter icon] *DIMDIAmeter* Diameter	Use the Dimension Element tool to place a quick diameter on a circle or arc. Use the tool setting Dimension Diameter to control whether you get a radius or a diameter. Use the Diameter or Extended Diameter tool to place a diameter on a circle or arc.

AutoCAD	MicroStation
Mtext	There is no equivalent option in MicroStation. You can edit the text in the dimension after placement.
Text	There is no equivalent option in MicroStation. You can edit the text in the dimension after placement.
Angle	There is no exact equivalent option in MicroStation. You can rotate the dimension text to vertical using the Text Alignment tool setting.

DIAMETER PERPENDICULAR

The Diameter Perpendicular tool will dimension the diameter perpendicular to the circle with an interior perpendicular dimension.

1 Select the Dimension Diameter Perpendicular tool.

2 Select the dimension style to be used.

3 Select the circular element to be dimensioned.

4 Issue a data point for the dimension location.

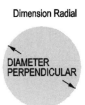

Dimension Radial

DIAMETER PERPENDICULAR

AutoCAD Command Comparison

There is no exact equivalent command in AutoCAD for the Diameter Perpendicular tool.

DIAMETER PARALLEL

The Diameter Parallel tool will dimension the diameter parallel of a circle axis with a linear dimension.

1 Select the Dimension Diameter Perpendicular tool.

2 Select the dimension style to be used.

3 Select the circular element to be dimensioned.

4 Issue a data point for the dimension location.

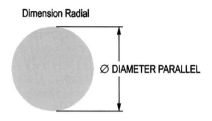

AutoCAD Command Comparison

AutoCAD	MicroStation
Mouse Pick: Keyboard: DIMLINea Linear	Use the Dimension Element tool to place a linear dimension on a radial element. Use the Diameter Parallel tool to place a linear dimension on a radial element.
Mtext	There is no equivalent option in MicroStation. You can edit the text in the dimension after placement.
Text	There is no equivalent option in MicroStation. You can edit the text in the dimension after placement.
Angle	There is no exact equivalent option in MicroStation. You can rotate the dimension text to vertical using the Text Alignment tool setting.
Horizontal	Drag the dimension using the mouse to control dimension axis.
Vertical	Drag the dimension using the mouse to control dimension axis.
Rotated	Use the Dimension Linear tool to place a rotated linear dimension. Control the angle using snap points on the dimensioned element.

DIMENSION RADIUS/DIAMETER NOTE

The Dimension Radius/Diameter Note tool will dimension the radius or diameter of a circle or arc and leave a leader line so that you can add a note.

1 Select the Dimension Radius/Diameter Note tool.

2 Select the dimension style to be used.

3 Select the type radius or diameter.

4 Select the circular element to be dimensioned.

5 Issue a data point for the dimension location.

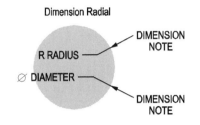

Dimension Radial

AutoCAD Command Comparison

There is no equivalent command in AutoCAD for the Dimension Radius/Diameter Note tool.

DIMENSION CENTER
The Dimension Center tool will dimension the center of a circle or arc.

1 Select the Dimension Center tool.

2 Select the dimension style to be used.

3 Define the center mark size.

4 Select the circular element to be dimensioned.

5 Issue a data point for the dimension location.

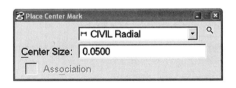

Dimension Radial

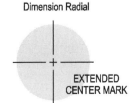

Dimension Radial

TIP: *Use a negative value for the center size to get the extended center mark lines.*

AutoCAD Command Comparison

AutoCAD	MicroStation
Mouse Pick: Keyboard: Center Mark *DIMCENTER*	Use the Dimension Element tool to place a linear dimension on a radial element. Use the Diameter Parallel tool to place a linear dimension on a radial element.
DIMCEN	Use the tool setting Center Size to define the size of the center mark.
Mtext	There is no equivalent option in MicroStation. You can edit the text in the dimension after placement.

DIMENSION ARC DISTANCE

The Dimension Arc Distance tool will dimension the perpendicular distance between arcs.

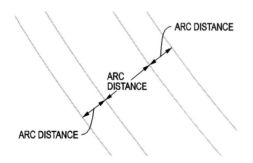

1   Select the Dimension Arc Distance tool.

2   Select the dimension style to be used.

3   Select the preferred alignment method.

4   Select the first arc element.

5   Select the second arc element.

6   Issue a data point for the dimension location.

AutoCAD Command Comparison

There is no equivalent command in AutoCAD for the Dimension Arc Distance tool.

CHANGE DIMENSION

Use the Change Dimension tool to modify the dimension settings of an existing dimension element to the active dimension settings.

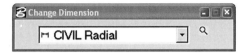

1   Select the Change Dimension tool.

2   Select the dimension style to be applied.

3   Select an existing dimension element.

4   Issue a data point to accept the existing dimension and apply the active dimension settings.

AutoCAD Command Comparison

AutoCAD	MicroStation
Mouse Pick: Keyboard: -*DIMSTYLE* Dimension Update	Use the Change Dimension tool to modify the dimension properties of an existing element.
Save	Use the **Element > Dimension Styles** command to save a new dimension style.
Restore	Use the Tool Settings dialog to restore a saved dimension style. Select the saved style and click on No to save changes.
Status	Use the **Element > Dimension Styles** command to review the status of a dimension style.
Variables	Use the Element Information tool to view the settings of an existing dimension.
Apply	Use the Change Dimension tool to apply the active dimension settings to an existing dimension.
?	Use the Tool Settings dialog to view available dimension styles while in a dimension command. Use the **Element > Dimension Styles** command to view available dimension styles.

AutoCAD Command Comparison

AutoCAD		MicroStation
Mouse Pick: Properties	Keyboard: *PRoperties*	Use the Change Dimension tool to modify the dimension properties of an existing element.

MATCH DIMENSION

Use the Match Dimension tool to match all text attributes from an existing text element.

1   Select the Match Dimension tool.

2   Identify an existing dimension element that looks correct.

3   Issue a data point to accept and apply dimension attributes to active settings.

AutoCAD Command Comparison

AutoCAD		MicroStation
Mouse Pick: Match Properties	Keyboard: *MAtchprop*	Use the Match Dimension tool to set the active dimension properties to those of an existing element. Use the SmartMatch tool to set all element attributes to those of an existing element.

REASSOCIATE DIMENSION

Use the Reassociate Dimension tool to fix a dimension that has become disassociated with elements.

Before you can reassociate a dimension you must fix the graphics so that the dimensions line up with their respective graphical elements. You cannot reassociate a dimension when the extension lines do not align with an element.

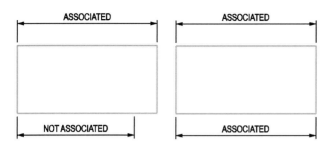

1   Select the Reassociate Dimension tool.

2   Select the dimension to be reassociated.

TIP: *You can use a selection set or a fence to reassociate multiple dimensions.*

AutoCAD Command Comparison

AutoCAD	MicroStation
Keyboard: *DIMREASSOCIATE*	Use the Reassociate Dimension tool to link dimensions back to their respective graphic elements.
Keyboard: *DIMDISASSOCIATE*	Use the Drop Association tool to remove the link between a dimension and another element.

DIMENSION STYLES

Dimension styles are similar to text styles, with the exception of their "save dimension element" settings. You can store these styles in a dimension style library to gain all of the benefits described for text style libraries. Dimension styles can be standardized by using a *.DGNLIB* file.

Non–Scale-based Dimension Styles

This type of dimension style defines the basic settings for the dimension elements but does not define text and terminator sizes for specific output scales. A style defined in this way would be used as a generic base for all dimension elements. However, you would be required to size the dimension text manually or with annotation scale during placement.

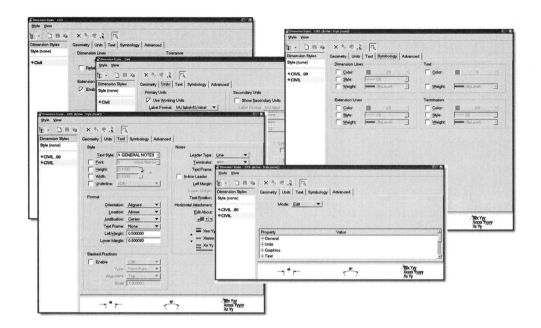

TIP: *Settings in blue have not been saved back to the style and can be used as a temporary override. Save the style to make temporary overrides permanent, or reset the style to discard these temporary overrides.*

Scale-based Dimension Styles

This type of dimension style defines the basic settings for dimension elements and the exact text and terminator sizes for specific output scales. A style defined in this way would be used as an automated method for placing dimension elements, and you would not be required to size the dimension text during placement. All dimension text sizes are saved in the actual dimension or text styles.

You can use text styles to define dimension text settings. This allows you to manage the text and dimension text settings within text styles and manage dimension terminators and graphics within the dimension styles. There are no parent/child relationships for dimension styles.

TIP: *There is a new dialog available for dimension styles, which is similar to the Text Style dialog. It is not loaded by default, and you must key in MDL L DIMSTYLE to load the application and then key in DIALOG DIMSTYLE to access this alternative dialog.*

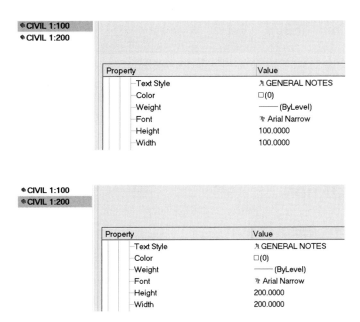

AutoCAD Command Comparison

AutoCAD	MicroStation
Format > Dimension Styles Keyboard: *DIMSTYLE*	Use the **Element > Dimension Styles** command to create and define dimension styles.

Dimension Audit

The Dimension Audit tool searches a model for invalid or incorrect dimensions and allows you to fix the problems as needed. This utility will search for the following common dimension problems.

❏ Overwritten text

❏ Dropped dimensions

❏ Non-associative dimensions

❏ Lost associations

AutoCAD Command Comparison	
AutoCAD Express Tool	MicroStation
Express > Dimension > Reset Dim Text Value Keyboard: *DIMREASSOC*	Use the **Utilities > Dimension Audit** command to reassociate lost and nonassociated dimensions.

ASSOCIATION

Element Association

Using association allows you to "link" elements in relationships determined by you. Normally, elements move freely of other elements unless they are cells, groups, or reference files. In most cases this is probably preferred. However, occasionally you need to "associate" a specific element with another. The elements that can be associated are cells, dimensions, and multi-lines. For example, you can associate a cell with another element, or you can associate a multi-line with another element.

File Association

You can associate files with specific extensions using the File Association dialog. There are primarily two methods of associating files: the drag-and-drop method and the link method. This is important to mention because a common practice in AutoCAD is to drag and drop files, graphics, and data information. If you do this in MicroStation, the response can be significantly different.

Be prepared for this difference in behavior when working with drag-and-drop techniques. Further details are provided throughout this book where drag-and-drop techniques available in MicroStation are discussed.

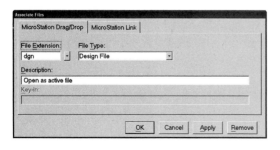
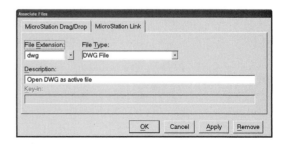

Annotation Scale

Annotation scaling is a new feature in V8 that provides the ability to define commonly used scales to control annotation elements automatically. Annotation scale is not available for all element types but is available for text and dimensions. This feature will probably be added to other element types in future releases of MicroStation.

The first requirement for using the annotation scale feature is to define the model scale which is used by annotation scale to determine text and dimensions sizes. The Drawing Scale dialog provides easy access to the current model scale settings. This dialog is available only through the following key-in.

```
dialog drawingscale open
```

This tool allows you to change the model scale to a preferred output scale for that design. This automatically scales all text and dimensions to the correct size for that output scale using the annotation scale factor.

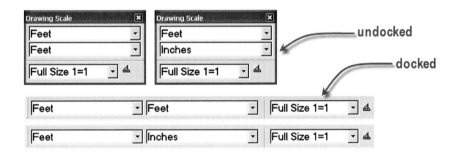

Text placed without using annotation scale will not be scaled automatically. You can add and remove the annotation scale factor to single or multiple elements using the following key-ins.

```
annotationscale add
annotationscale remove
annotationscale change
```

AutoCAD Command Comparison

AutoCAD	MicroStation
There is no exact equivalent command in AutoCAD. Viewport Scale can control the size of some text but it is not automatic.	Annotation scale.
Mouse Pick: Keyboard: `1:50` *Zoom Scale valueXP* Viewport Scale Control	Use the Annotation Scale tool to control the size of text and dimension elements based on the model scale.

In Exercise 7-1, following, you have the opportunity to practice working with text in a civil engineering context.

EXERCISE 7-1: WORKING WITH TEXT IN A CIVIL ENGINEERING ENVIRONMENT

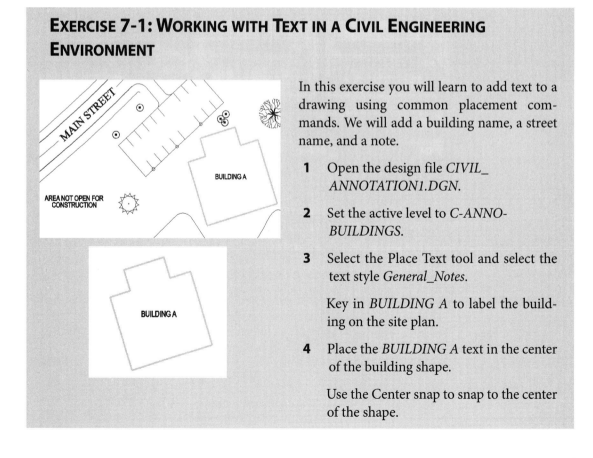

In this exercise you will learn to add text to a drawing using common placement commands. We will add a building name, a street name, and a note.

1 Open the design file *CIVIL_ ANNOTATION1.DGN*.

2 Set the active level to *C-ANNO-BUILDINGS*.

3 Select the Place Text tool and select the text style *General_Notes*.

Key in *BUILDING A* to label the building on the site plan.

4 Place the *BUILDING A* text in the center of the building shape.

Use the Center snap to snap to the center of the shape.

5 Set the active level to *C-ANNO-TEXT*.

6 Set the Method tool setting to WORD WRAP, and specify the window area for the text to be placed within.

Pick in the text editor dialog and key in the text *AREA NOT OPEN FOR CONSTRUC-TION*.

Issue a data point anywhere in the view window to complete the command.

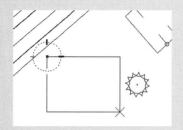 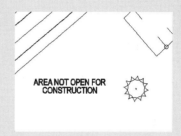

7 Set the active level to *C-ANNO-STREET_NAMES*.

8 Set the Method tool setting to ON ELEMENT.

Set the Text Style tool setting to STREET_NAMES.

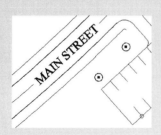

9 Key in the street name *MAIN STREET*.

10 Select the street centerline to place the *MAIN STREET* text on the main road located at the top of view 1.

11 Issue a data point to accept the text location.

You can issue a reset on the mouse to select a new text location if needed

In Exercise 7-2, following, you have the opportunity to practice working with dimensions in a civil engineering context.

EXERCISE 7-2: WORKING WITH DIMENSIONS IN A CIVIL ENGINEERING ENVIRONMENT

In this exercise you will learn to add dimensions to a drawing using common placement commands. We will add the building dimensions.

1 Open the design file *CIVIL_ANNOTATION1.DGN*.

2 Set the active level to *C-ANNO-DIMS*.

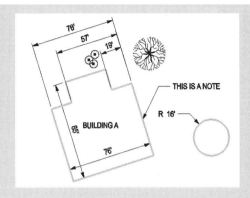

3 Select the Dimension Element tool and select the south edge of the building outline.

Drag the dimension into position and issue a data point to accept this location.

4 Select the Dimension Linear tool to place the dimension on the west side of the building.

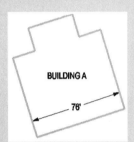

5 Using the keypoint method, snap to P1 for the start of the dimension.

Press the Enter key to activate the AccuDraw SmartLock function along the X axis.

6 Using the keypoint method, snap to P2 for the dimension end point.

Drag the dimension into position and issue a data point to accept the location.

Reset to end the dimension command.

7 Select the Select Multiple Elements button in the Tool Settings dialog.

8 Using the keypoint method, snap to P3 for the start of the dimension.

You will need to use a tentative point to snap to P3.

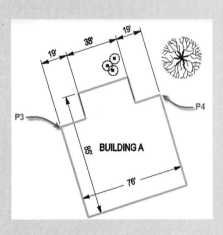

9 Using another tentative point, snap to the keypoint at P4.

Drag the dimension string into position and issue a data point to accept the location.

10 Select the Place Note tool and key in the text *THIS IS A NOTE.*

11 Snap to the east side of the building outline and drag the note into position.

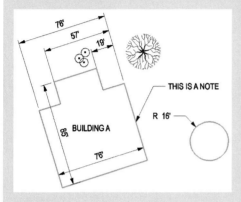

12 Turn on the In-line Leader tool setting if you want a straight line on the end of the leader line.

Issue a data point to accept the location.

13 Select the Dimension Element tool to dimension the radius of the round tank.

Set the Dimension Style setting to CIVIL RADIAL in the Tool Settings dialog.

Select the tank circle and drag the dimension into position and issue a data point to accept the location.

14 Delete the tank element and note the change in the radial dimension appearance.

Heavy dashed elements represent a "broken association."

15 Draw another circle in the same location as the old tank. Be sure to place the new circle center at the point displayed and snap to the arrow on the leader dimension.

16 Use the Reassociate Dimension tool and select the radial dimension to reassociate it to the circle.

8:   Productivity Using Cells

CHAPTER OBJECTIVES:

❑   Learn the basics of using cells

❑   Learn how to create cells

❑   Learn to use AutoCAD blocks

❑   Learn to use cell libraries and cell utilities

Using standard symbols is an important part of any drawing package. The concept of using symbols is identical in both CAD packages. In MicroStation, standard symbols are called cells, not blocks, and they are stored in cell libraries. Typical MicroStation cells are created somewhat differently in regard to level, color, line style, and line weight.

TYPES OF CELLS

This section explores the various types of cells available and their differences. MicroStation cells have several options available, depending on how you want to define your CAD standards. There are three basic types of cells found in MicroStation: graphic cells, point cells, and menu cells.

Graphic

A graphic cell type is a symbol for which the following are true.

❑   Its element symbology (color, line style, and line weight) is determined when it is created, not when it is placed.

❑   Its level is independent. It can be placed "absolute" on the levels defined in the cell, or "relative" to the levels defined in the cell.

❑ It has multiple snappable points.

❑ It rotates with a view.

Point

A point cell type is a symbol for which the following are true.

❑ Its element symbology (color, line style, and line weight) is determined when it is placed, not when it is created.

❑ It is level dependent. It is placed on the active level.

❑ It has one snap point located at the "origin" of the cell.

❑ It is view independent and will not rotate with a view. This is very useful when a symbol contains text that needs to remain right-reading.

Menu

The menu cell type is virtually obsolete, but can still be used for generating tablet menus or screen menus.

USING CELLS

MicroStation J

Using MicroStation J, graphic elements for symbols are typically placed on the correct level prior to making the cell. The color, line style, and line weight should also be applied to the elements before they are added to a cell. Using this technique saves time later, and you do not have to worry about the active attributes when the cell is placed. The cell already knows what attributes you need; they are built into the cell itself. Cells containing multiple levels, colors, line styles, and line weights are considered "normal."

MicroStation V8

A new option is available for using cells in MicroStation V8. The graphic elements for the cell can be placed on the "default" level, and using ByLevel or ByCell you can obtain more control over the placement symbology than was previously available using a point cell. Point cells are still available, and still have a place in the DGN file. However, ByLevel cells offer some enhancements. ByLevel cells can contain multiple colors, line styles, and line weights. The following should be noted.

❑ A cell created on level "default" works exactly like the "0" level in AutoCAD. It is a generic level designation that can be used to inherit the active level during placement.

❑ A cell created using ByLevel for color, line style, and line weight can be used to inherit the level definition symbology during placement.

❑ A cell created using ByCell for color, line style, and line weight can be used to inherit the level definition or active symbology during placement.

AutoCAD

You can even use AutoCAD blocks in their original DWG file format as cells in MicroStation. Most AutoCAD blocks work in a manner similar to that of ByLevel cells in V8.

CELL LIBRARIES

Cell libraries are very similar to the collections of external WBlock files used in AutoCAD, with one major exception. WBlock files generally contain only one symbol per file, whereas cell libraries contain many symbols in a single file. You can place all symbols in a single cell library or you can create discipline-specific cell libraries and divide the symbols between them. The second option is more useful, in that you do not want to sort through thousands of standard symbols looking for a specific manhole or annotation symbol. It would be more organized and efficient to have separate libraries for discipline-specific symbols.

BLOCK LIBRARIES

AutoCAD blocks are generally organized into folders containing the individual symbol files (called WBlock files). These files can be directly accessed using the configuration settings available in the MicroStation workspace. The following is the MicroStation configuration setting that defines the location of these WBlock files.

```
MS_BLOCKLIST = WBlock folder locations
```

AutoCAD blocks organized into DWG files with multiple symbols in a single file are not accessible from within MicroStation. These multiple block storage files should be "split up" and organized using the individual WBlock file method discussed previously.

LEARNING THE LANGUAGE

Table 8-1 outlines MicroStation and AutoCAD terminology equivalents.

TABLE 8-1: MICROSTATION AND AUTOCAD TERMINOLOGY EQUIVALENTS

AutoCAD	MicroStation
Blocks	Cells
Insert	Place
Insertion point	Origin
Hatch	Pattern cell
ByLayer	ByLevel
ByBlock	ByCell
Tool Palette	
Design Center	
WBlock	

CREATING CELLS

The process used to create a cell in MicroStation is almost identical to that in AutoCAD.

1 Draw the graphic components of the cell.

2 Define the origin or insertion point.

3 Name the cell or block.

4 Store the cell or block.

When these steps have been completed, the symbol is ready to be used.

STORING CELLS

This is where the two applications differ.

AutoCAD

AutoCAD typically stores blocks in separate files called WBlock files. These files are simple DWG files stored in an organized folder structure on a server. Some AutoCAD users are beginning to store blocks in a single file on a server using utilities such as Design Center or tool palettes to access them. Blocks stored via this method are not easily accessed by MicroStation.

MicroStation

MicroStation stores cells in cell libraries. A cell library has the identical format as a DGN file in V8. However, older cell libraries are not. You have to convert V7 libraries to V8 to open and edit cell libraries easily.

Create a Cell Library

Creating a cell library is as easy as creating a DGN file. You can access the Create Cell Library tool from the Cell Library dialog. This dialog also provides a user-friendly interface for most cell commands.

PLACE CELL
Use the Place Cell tool to place standard symbols in your drawing. Cells are the equivalent of blocks in AutoCAD and provide easy access to standardized symbols through the use of cell libraries. There are several tool settings that can affect how you place cells. These settings affect how absolute and relative cells are controlled.

ACTIVE CELL

The Active Cell tool controls what symbol will be placed. Use the Find Cell button to open the Cell Library dialog and search for cells in the attached library.

ACTIVE ANGLE

The Active Angle tool defines the angle for placing cells. Use the navigation arrows to scroll between commonly used angles.

X SCALE

The *X scale* tool defines the horizontal scale factor for placing cells. Use the Lock icon to lock the X and Y scales together for proportional cells.

Y SCALE

The *Y scale* tool defines the vertical scale factor for placing cells.

TRUE SCALE

The True Scale setting forces cells to scale automatically when used between files of different working units. Metric cells and Imperial cells can be used interchangeably without concern for the scale factors required to match the units equally. Thus, a single cell library can contain both metric and Imperial cells.

RELATIVE

Using this setting allows you to modify the level a cell is placed on. Normally a cell is placed on the level it was created on (called an "absolute" cell). With this setting you can change the level the cell resides on during placement by placing it as a cell "relative" to the active level.

INTERACTIVE

This setting allows you to graphically define the scale and angle during cell placement.

FLATTEN

This setting allows you to "flatten" a 3D cell for use in 3D or 2D files. You can specify which view of the 3D cell you want to place in its "flattened" state.

ASSOCIATION

The Association setting allows you to associate or link a cell to another graphic element.

AutoCAD Command Comparison

AutoCAD	MicroStation
Mouse pick: Keyboard: *Insert* Insert Block	Use the Place Cell tool and define the cell name, rotation angle, and scale in the Tool Settings dialog.
Insertion Point	Issue a data point to specify the location of the cell in the drawing.
Rotation	Use the Active Angle tool setting.
Scale X,Y, Z	Use the *X scale*, *Y scale*, or *Z scale* tool settings to define the cell scale.
Basepoint	Not available within the Place Cell tool. Move the cell after placement.

WORKING WITH CELLS

SELECT AND PLACE CELL

Use the Select and Place Cell tool to select an existing cell and place it again easily without having to know where it is stored or what it is named. The existing cell tool settings such as scale, rotation, and attributes are not recalled during this placement command.

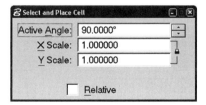

AutoCAD Command Comparison

There is no equivalent command in AutoCAD for the Select and Place Cell tool.

DEFINE CELL ORIGIN

Use the Define Cell Origin tool to define the insertion point of a cell during the creation process. The cell origin is represented as a "white crosshair" on the view window. This on-screen origin point can only be removed from the view window by selecting the Define Cell Origin tool again.

Note that the cell origin is not treated as an element for any purpose other than view operations. It is intended as a visual reminder of the cell origin location. Select the Cell Origin tool again to clear the cell origin from the view.

AutoCAD Command Comparison

AutoCAD	MicroStation
Mouse pick: Keyboard: *Block* Make Block	Use the Define Cell Origin tool to define the cell base point during the cell creation process. Use the Create Cell command from within the Cell Library dialog to create cells. Be sure to select the cell elements to enable the Create button.
Insertion Point	Issue a data point to specify the location of the cell in the drawing.
Rotation	Use the Active Angle tool setting.
Scale X,Y,Z	Use the *X scale*, *Y scale*, or *Z scale* tool settings to define the cell scale.
Basepoint	Not available within the Place Cell tool. Move the cell after placement.

IDENTIFY CELL

Use the Identify Cell tool to identify an existing cell and display its name and residing level. This information is displayed in the Message Center located in the status bar at the bottom of the application window.

AutoCAD Command Comparison

AutoCAD		MicroStation
Mouse pick: Properties	Keyboard: *PRoperties*	Use the Identify Cell tool to identify cell name and level. Use the Element Information tool for all other cell information.

PLACE LINE TERMINATOR

Use the Place Line Terminator tool to place cells automatically aligned with existing graphics. This is especially useful with directional symbols such as arrowheads and flow arrows.

AutoCAD Command Comparison

There is no equivalent command in AutoCAD for Place Line Terminator tool.

REPLACE CELLS

Use the Replace Cells tool to update existing cells to newer versions of the same cell or to replace cells with different cells. These replacements can be performed individually or globally in a design file. The following cell settings can be modified.

❑ Cell definition

❑ True Scale activation

❑ Tag data

❑ Element attributes such as color, line style, or line weight

❑ Levels

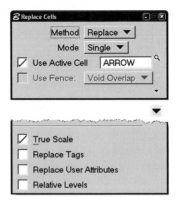

AutoCAD Command Comparison

AutoCAD Express Tool	MicroStation
Express > Blocks > Replace Block with another block Keyboard: *BLOCKREPLACE*	Use the Replace Cells tool to replace existing cells with an updated or new cell.

In Exercise 8-1, following, you have the opportunity to practice using the graphic cell type.

EXERCISE 8-1: USING A GRAPHIC CELL

In this exercise you will learn to place graphic cells in a drawing file. Cells are placed using the active tool settings to control angle and scale.

Absolute Graphic Cell

1　Open the design file *GRAPHIC_CELLS.DGN*.

2　To attach a cell library, go to the pull-down menu **Element > Cells**.

3　This will open the Cell Library dialog. From this dialog you can view the cells contained in the attached cell library *Common_CELLS.CEL*.

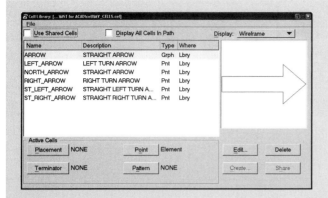

4 Select the cell *ARROW* from the list to view the cell in the preview window.

The cell *ARROW* is a graphic cell.

5 Click on the Placement button to set the *ARROW* cell as the active placement cell.

6 Select the Place Cell tool from Main toolframe > Cells toolbar.

Place the cell by issuing a data point anywhere in the drawing. Verify the following tool settings.

Active Angle: 0.0000
X scale: 1.000000
Y scale: 1.000000

The active attributes are:

Level: default
Color: 2 (green)
Line Style: 4 (dashed dot)
Line Weight: 2

The placed cell's attributes are:

Level: as created   level 1
Color: as created   5 (magenta)
Line Style: as created   0 (continuous)
Line Weight: as created   2

7 Place the cell again using the following tool settings.

Active Angle: 90.0000
X scale: 2.000000
Y scale: 2.000000

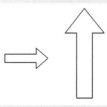

This cell has the identical attributes, but points in the 90-degree direction and is twice the size of the previous cell placed.

Relative Graphic Cell

Let's take a look at some of the options available for placing cells such as relative, interactive, and associated. Let's place the same cell using the Relative tool setting. Placing a graphic cell relative places the cell on the active level instead of the level it was created on. It does not affect the color, line style, or line weight attributes.

8   Enable the "hidden" tool setting options and activate the Relative tool setting.

The active attributes are:

Level:	default
Color:	2 (green)
Line Style:	4 (dashed dot)
Line Weight:	2

The placed cell's attributes are:

Level:	**active level**	**default**
Color:	as created	5 (magenta)
Line Style:	as created	0 (continuous)
Line Weight:	as created	2

Interactive Graphic Cell

Now let's place the cell using the Interactive tool settings. Placing a cell interactive allows you to graphically define the scale and rotation of the cell. It has no impact on the cell's symbology attributes.

9   Deactivate the Relative tool setting.

10   Activate the Interactive tool setting and place the cell *ARROW* again.

11   Issue a data point in the view window to specify the insertion point of the cell.

12   Move your cursor in the NW direction (↖) to graphically define the X and Y scale.

Issue a data point to accept the scale when the cell looks the way you need.

13   Move the cursor in a circular motion to graphically define the rotation angle you need.

Issue a data point to accept the rotation angle when the cell looks the way you need.

Let's place the same cell using the Association tool setting. Placing a cell using an association allows you to link that cell to another element in the drawing. It has no impact on the cell's symbology attributes.

14 Deactivate the Interactive tool setting.

15 Activate the Association tool setting and snap to the midpoint of the existing rectangle to place the cell.

Use the Move Element tool to move the rectangle. The *ARROW* cell will follow the rectangle automatically because it is associated with the rectangle.

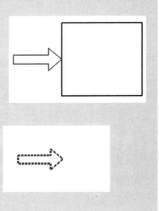

16 Delete the rectangle using the Delete tool.

Note that the *ARROW* cell is now displayed using the "broken" association symbology.

17 Use the Undo option to undo the last delete command and to restore the cell association.

In Exercise 8-2, following, you can practice using the point cell type.

EXERCISE 8-2: USING A POINT CELL

In this exercise you will learn to place point cells in a drawing file. Cells are placed using the active tool settings to control angle and scale.

Absolute Point Cell

1 Open the design file *POINT_CELLS.DGN*.

2 Select the cell *ARROW_P* from the list of cells in the Cell Library dialog and view the cell in the preview window.

The cell *ARROW_P* is a point cell.

3 You can double click on the cell in the list to set the *ARROW_P* cell as the active placement cell.

This will also automatically run the Place Cell command, saving you time and mouse picks.

4 Place the cell by issuing a data point anywhere in the drawing using the following tool settings.

Active Angle: 0.0000
X scale: 1.000000
Y scale: 1.000000

The active attributes are:

Level: level 1
Color: 2 (green)
Line Style: 4 (dashed dot)
Line Weight: 2

The placed cell's attributes are:

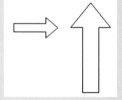

Level: active level level 1
Color: active color 2 (green)
Line Style: active style 4 (dashed dot)
Line Weight: active weight 2

Relative Point Cell

Placing a point cell using the Relative tool setting is redundant because a point cell automatically inherits all active attributes.

Interactive Point Cell

Placing a point cell using the Interactive tool settings is identical to placing an absolute graphic cell because the Interactive option has no impact on the cell's symbology.

Association Point Cell

Placing a point cell using the Association tool settings is identical to placing an absolute graphic cell because the association option has no impact on the cell's symbology.

In Exercise 8-3, following, you have the opportunity to practice using the ByLevel cell.

EXERCISE 8-3: USING A BYLEVEL CELL

In this exercise you will learn to use cells in an "AutoCAD-like" environment using the new Default level available in V8. You will also learn the effects of using the ByLevel and ByCell attribute settings in MicroStation cells.

Using the Default Level in a Cell

Using the Default level in a cell definition causes the cell to be placed in a "relative" mode regardless of the tool setting options. Using the Relative tool setting will have no effect on this cell.

1 Open the design file *BYLEVEL_CELLS.DGN.*

2 To place the cell *ARROW_Default* double click on the cell name to set it active and to activate the Place Cell tool.

Verify the following tool settings.

Active Angle: 0.0000
X scale: 1.000000
Y scale: 1.000000

The active attributes are:

Level: level 1
Color: 2 (green)
Line Style: 4 (dashed dot)
Line Weight: 2

The placed cell's attributes are:

Level:	**active level**	**level 1**
Color:	as created	5 (magenta)
Line Style:	as created	0 (continuous)
Line Weight:	as created	2

Using the ByLevel Attribute in Cells

Using a cell with the color defined as ByLevel allows you to control the color of the cell using the level definition, not the active color or created color. The level for this cell is set to Default so that the cell will inherit the active level for demonstration purposes.

3 Change the active level to "blue dashed" and place the cell *ARROW_bylevel_color.*

The active attributes are:

Level: blue dashed
Color: 2 (green)
Line Style: 4 (dashed dot)
Line Weight: 2

The placed cell's attributes are:

Level:	active level	blue dashed
Color:	**bylevel color**	**1 (blue)**
Line Style:	as created	0 (continuous)
Line Weight:	as created	2

4   Change the active level to "red thin" and place the cell again.

The active attributes are:

Level:	red thin
Color:	2 (green)
Line Style:	4 (dashed dot)
Line Weight:	2

The placed cell's attributes are:

Level:	active level	red thin
Color:	**bylevel color**	**3 (red)**
Line Style:	as created	0 (continuous)
Line Weight:	as created	2

Using ByLevel for the color allows you to use a symbol on more than one level and adheres to the color standards by inheriting the color from the standard level.

Next, let's place a cell using all ByLevel attributes for color, line style, and line weight. The level for this cell is set to Default so that the cell will inherit the active level for demonstration purposes.

5   Change the active level to "blue dashed" and place the cell *ARROW_bylevel_all.*

The active attributes are:

Level:	blue dashed
Color:	2 (green)
Line Style:	4 (dashed dot)
Line Weight:	2

The placed cell's attributes are:

Level:	**active level**	**blue dashed**
Color:	**bylevel color**	**1 (blue)**
Line Style:	**bylevel style**	**2 (dashed)**
Line Weight:	**bylevel weight**	**1**

6   Change the active level to "green thick" and place the cell *ARROW_bylevel_all.*

The placed cell's attributes are:

Level:	**active level**	**green thick**
Color:	**bylevel color**	**2 (green)**
Line Style:	**bylevel style**	**0 (continuous)**
Line Weight:	**bylevel weight**	**2**

7   Change the active level to "red thin" and place the cell *ARROW_bylevel_all.*

The placed cell's attributes are:

Level:	**active level**	**red thin**
Color:	**bylevel color**	**3 (red)**
Line Style:	**bylevel style**	**0 (continuous)**
Line Weight:	**bylevel weight**	**0**

Using the ByLevel attribute setting eliminates the effects of active attributes for color, line style, and line weight. They no longer have any impact on the cell's symbology. This is very effective in enforcing standard symbology settings associated with standard levels.

However, this can cause problems if you want more than one color, line style, or line weight in a single cell.

Next, let's place a cell using generic ByLevel attributes and hard-coded assigned attributes for color, line style, and line weight.

8   Change the active level to "blue dashed" and place the cell *ARROW_partial_bylevels.*

The active attributes are:

Level:	blue dashed
Color:	2 (green)
Line Style:	4 (dashed dot)
Line Weight:	2

The placed cell's attributes are:

Level:	**active level**	**blue dashed**
Color:	**bylevel color**	**1 (blue)**
Line Style:	**bylevel style**	**2 (dashed)**
Line Weight:	**bylevel weight**	**1**

9   Change the active level to "green thick" and place the cell *ARROW_ bylevel_all.*

The placed cell's attributes are:

Level:	**active level**	**green thick**
Color:	**bylevel color**	**2 (green)**
Line Style:	**bylevel style**	**0 (continuous)**
Line Weight:	**bylevel weight**	**4**

10 Change the active level to "red thin" and place the cell *ARROW_bylevel_all*.

The placed cell's attributes are:

Level:	**active level**	**red thin**
Color:	**bylevel color**	**3 (red)**
Line Style:	**bylevel style**	**0 (continuous)**
Line Weight:	**bylevel weight**	**0**

Using the ByCell Attribute in Cells

Using a cell with the color defined as ByCell allows you to control the color of the cell using the active attributes, rather than the level definition color or cell definition color. Cells using a ByCell attribute must be placed as shared cells.

11 Open the Cell Library dialog to activate the Shared Cells placement option.

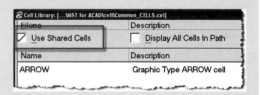

12 Place the cell *ARROW_bycell_color*.

The active attributes are:

Level:	blue dashed
Color:	2 (green)
Line Style:	4 (dashed dot)
Line Weight:	2

The placed cell's attributes are:

Level:	active level	blue dashed
Color:	**active color**	**2 (green)**
Line Style:	bylevel style	2 (dashed)
Line Weight:	bylevel weight	2

13 Change the active color to 6 (orange) and place the cell again.

The active attributes are:

Level:	blue dashed
Color:	6 (orange)
Line Style:	4 (dashed dot)
Line Weight:	2

The placed cell's attributes are:

Level:	active level	blue dashed
Color:	**active color**	**6 (orange)**
Line Style:	bylevel style	2 (dashed)
Line Weight:	bylevel weight	2

Next, let's place a cell using all ByCell attributes for color, line style, and line weight.

14 Change the active level to "blue dashed" and set all other active attributes to ByLevel.

15 Place the cell *ARROW_bycell_all*.

The active attributes are:

Level:	blue dashed
Color:	bylevel (blue)
Line Style:	bylevel (dashed)
Line Weight:	bylevel (2)

The placed cell's attributes are:

Level:	**active level**	**blue dashed**
Color:	**bylevel color**	**1 (blue)**
Line Style:	**bylevel style**	**2 (dashed)**
Line Weight:	**bylevel weight**	**2**

Note that the cell works exactly like its BYLEVEL counterpart when all active attributes are set to ByLevel.

16 Change the active color to 2 (orange) and place the cell again.

The modified placed cell's attribute is:

Color:	**active color** **2 (orange)**

17 Change the active line style to 0 (continuous) and place the cell again.

The modified placed cell's attribute is:

Line Style:	**active style** **0 (continuous)**

18 Change the active line weight to 0 (thin) and place the cell again.

The modified placed cell's attribute is:

Line Weight:	**active weight** **0 (thin)**

This type of cell gives you the most flexibility when it comes to generic symbols.

In Exercise 8-4, following, you have the opportunity to practice using AutoCAD blocks.

EXERCISE 8-4: USING AUTOCAD BLOCKS

In this exercise we will use AutoCAD blocks as standard symbols for MicroStation. You do not have to convert your AutoCAD blocks to cells unless you want to take advantage of alternative functionality available in MicroStation.

Using ByLayer Blocks

AutoCAD's ByLayer blocks work in a manner similar to that of ByLevel cells in MicroStation.

1   Open the design file *ACAD_BLOCKS.DGN*.

2   Open the Cell Library dialog to activate the Display All Cells in Path dialog option.

3   Place the cell *ACAD_bylayer block*.

The active attributes are:

Level:	blue dashed
Color:	2 (green)
Line Style:	4 (dashed dot)
Line Weight:	2

The placed cell's attributes are:

Level:	active level	blue dashed
Color:	**bylevel color**	**1 (blue)**
Line Style:	**bylevel style**	**2 (dashed)**
Line Weight:	**bylevel weight**	**2**

4   Change the active level to "red thin" and place the cell again.

The placed cell's attributes are:

Level:	**active level**	**red thin**
Color:	**bylevel color**	**3 (red)**
Line Style:	**bylevel style**	**0 (continuous)**
Line Weight:	**bylevel weight**	**0**

Using ByBlock Blocks

AutoCAD's ByBlock blocks work in a manner similar to that of ByCell blocks in MicroStation. Cells using a ByBlock attribute must be placed as shared cells.

5   Place the cell *ACAD_byblock block*.

The active attributes are:

Level: blue dashed
Color: 2 (green)
Line Style: 4 (dashed dot)
Line Weight: 2

The placed cell's attributes are:

Level: **active level** **blue dashed**
Color: **active color** **2 (green)**
Line Style: **active style** **4 (dashed dot)**
Line Weight: **active weight** **2**

MORE CELL PLACEMENT UTILITIES

Cell Selector

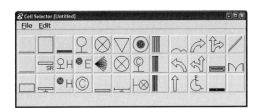

The Cell Selector utility provides an alternative interface for placing cells in the design file. Using the cell selector provides some additional control methods for placing cells. You can place the cell just as you do from the Cell Library dialog, but you can also add key-ins to set MicroStation attributes so that you can automatically define levels, colors, and so on. The cell selector also provides an easy-to-use dialog with pictures of all cells and helps automate commonly used symbols.

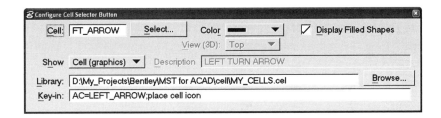

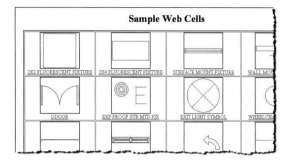

HTML Author

The HTML Author utility generates HTML pages for accessing standard symbols from the Web. Use this utility to generate web pages that provide Internet or intranet access to your standard symbols.

9: The Ins and Outs of Printing

CHAPTER OBJECTIVES:

❑ Learn the basics of plotting

❑ Learn how to configure and customize the plotting environment

❑ Learn to resymbolize design data through plotting

❑ Learn to batch your output needs using PDF output from within MicroStation

MicroStation provides printing capabilities similar to those found in AutoCAD and any other Windows application. However, MicroStation allows for additional controls using Bentley plot driver files, which you can easily edit and modify for your specific output needs. These controls include custom paper sizes, plot stamps, line weight controls, line style controls, and many others.

THE PRINT DIALOG

The Windows Print dialog provides access to the most commonly used settings for printing from MicroStation. You can expand the Print dialog to display additional print options using the "hidden" settings buttons provided.

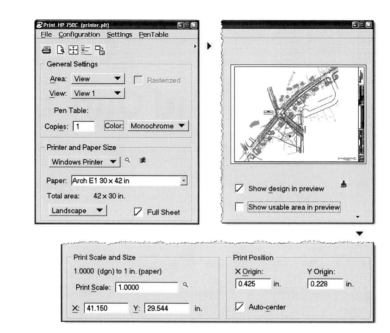

Several options in the Print dialog vary depending on the driver selected.

Windows Printer

Using a Windows driver offers options based on Windows-defined printers and driver settings. These drivers may contain too many limitations for some printer settings.

Bentley Driver

Using a Bentley driver offers options based on Bentley-defined printers and driver settings. These drivers offer more flexibility than Windows drivers for some plot devices.

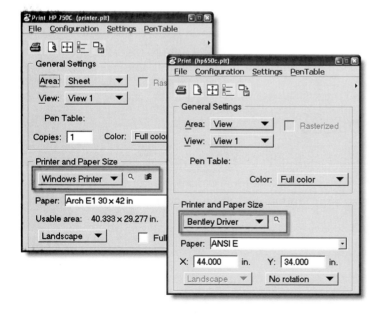

PRINT COMMANDS

There are several commands available from within the Print dialog.

Print

Use the Print tool to send the print to the output device.

Preview

Use the Preview tool to open a resizable preview window for a larger preview of the final output prior to executing the print command.

Maximize Print Size

Maximize Print Size maximizes the printable area defined by the Area setting in the available paper size.

Print Attributes

Print Attributes defines the view attributes used for printed output. These settings will override any settings defined in the selected area, such as view, sheet, and so on.

Update from View

Update from View allows you to refresh the preview window in the Print dialog with changes made to the view window after the Print dialog was opened.

General Settings

Use the following settings to specify general plot output parameters.

AREA
Area defines the extents of the drawing to be printed.

VIEW
View defines which view window and view levels and attributes will be printed.

PEN TABLE
Pen Table defines the pen table whose settings will be applied to the printed output.

COPIES
The Copies setting defines the number of copies to be printed. This option is available using the Windows Printer option only.

COLOR
Color defines the color depth of the printed output. The options are monochrome, grayscale, and color.

Printer and Paper Size Settings

Use the following settings to control the actual paper sizes and orientation.

WINDOWS PRINTER
Windows Printer uses the driver file *Printer.plt* to define the plot settings.

BENTLEY DRIVER
Use the specified Bentley driver file to define the plot settings.

PAPER
Select the paper size required from the selected driver file.

USEABLE AREA
Useable Area displays the actual printable area for the selected paper size and is available using the Windows Printer option only.

ORIENTATION OPTIONS
Orientation Options define the paper orientation to be used during output. The options of landscape and portrait are available using the Windows Printer option only.

FULL SHEET
Full Sheet controls the width and height for the printable area or full paper size.

ROTATION OPTIONS
Rotation Options define the desired output rotation if a different rotation is required from the current view orientation. This option is beneficial for large outputs that must be rotated to fit on actual paper sizes and to accommodate paper roll limitations. You can select the options No Rotation, Rotate 90 Clockwise, and Rotate 90 Counterclockwise.

AutoCAD Command Comparison

AutoCAD	MicroStation
Mouse pick: Keyboard: *PLOT* Plot	Use the settings in the Print dialog to plot drawing files.
Page Setups	You can define standard plot configurations using the Configuration pull-down menu found in the Printer dialog. Save your standard configurations as you do page setups in AutoCAD.
Printer/Plotter Name	Use the plot driver files to control the printer types.
Paper Size	Use the Paper setting found in the Printer and Paper Sizes section of the Print dialog. The available sizes are defined in the plot driver file.
Plot Area	Use the Area and View settings in the General section of the Print dialog.
Plot Scale	Use the Print Scale setting found in the Print Scale and Size section of the Print dialog. You might have to expand the bottom portion of the dialog to access this setting.
Plot Style Table	Use a pen table to remap and modify the appearance of the design file for plotting purposes only.
Plot Stamp	Use the Border option found in Print Attributes for a date and time stamp. Or you can use the Text Substitution feature found in the pen tables to place a plot stamp at any location on the output.
Portrait and Landscape	Use the Portrait and Landscape options found in the Printer and Paper Size section of the Print dialog. The availability of this setting is controlled by the plot driver selected.

SETTING UP YOUR FIRST PLOT

One of the first decisions is whether you are going to use Windows-defined printers or MicroStation-defined printers. Both are fully functional and the decision of which to use will be determined by the device drivers and their capabilities. Depending on the plot device you are using, the Windows drivers may not offer all of the options provided by the Bentley drivers.

Pen Tables

Use pen tables to modify the symbology between the design file and the printed output. Pen tables provide complete control over plotted output without having to modify the actual design file. You can define search criteria and output action criteria to completely change the appearance and content of a drawing. These pen tables are similar to the plot style tables available in AutoCAD. The more common symbology criteria include the following.

- ❑ Element mapping
- ❑ Screening
- ❑ Plot order
- ❑ Text substitution

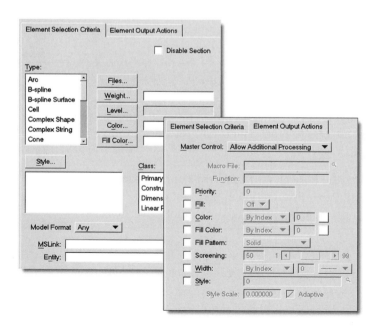

ELEMENT MAPPING

The Element Mapping output action allows you to modify output colors based on files, element type, line weight, level, color, fill color, models, and more. Once you specify the design file search criteria you can modify the output criteria using priority, fills, colors, patterns, screening, line widths, and line styles.

SCREENING

Use the Screening output capability to screen drawing components based on defined search criteria. You can set the screening output effect from 1 to 99 percent.

PLOT ORDER

Use Plot Order output action to control the priority order in which drawing components are plotted. You can define the search criteria based on files, element type, line weight, level, color, fill color, models, and so on. This is very beneficial when using filled elements or heavy line weight to avoid hidden graphics.

TEXT SUBSTITUTION

Use the Text Substitution output action to modify the content of specific text in the drawing for plotting purposes only. This is very similar to the remote text capabilities found in AutoCAD. You can use this to populate typical information in a border such as file names, pen tables, plot drivers, dates, user names, plot scales, paper sizes, model names, page numbers, and many other configuration values defined in the workspace environment.

IMPORT AUTOCAD PLOT STYLE TABLES

You can use your standard AutoCAD plot style tables as a starting point for MicroStation pen tables. Import your CTB or STB plot style table into a pen table to easily transfer the plot settings you are most familiar with.

EXPORT MICROSTATION PEN TABLE TO CTB

You can export a MicroStation pen table to an AutoCAD CTB file if needed.

AutoCAD C6mmand Comparison

AutoCAD	MicroStation
File > Plot Style Manager	Use pen tables to simulate the same behavior as plot style tables.
Line Weight	Use the element line weight by default, but you can change the plotted weight using level symbology or pen tables.
Screening	Use a pen table to plot using screened output.
Grayscale	Use a plot driver file or a pen table to plot using grayscaled output.

Using Bentley Device Drivers

There are several Bentley device drivers provided with MicroStation that allow you to customize and take advantage of large-format print capabilities. The Bentley drivers are provided in text format files so that they can be easily edited using any text editor. They can be found in the following location:

C:\Program Files\Bentley\Workspace\System\plotdrv

The following are just some of the settings you can control using a Bentley driver *.PLT* file.

❑ Available Colors

❑ Pen Table Definitions

❑ Page Setups and Sizes

❑ Line Weight Definitions

❑ Line Style Definitions

AVAILABLE COLORS

Edit the provided Bentley driver files and look for the following definition.

```
num_pens = 1     ; Monochrome output (black only)
num_pens = 255   ; Color or grayscale output
```

You can also control specific colors in your plotting output.

```
PEN(2)=(7)/rgb=(0,0,0)                 ; change white color to black
PEN(3)=(4,20,36,52,68,84,28,44,60,76,92)/rgb=(0,0,0)
                                       ; change yellow colors to black
PEN(4)=(8,9)/rgb=(128,128,128)   ; change colors 8 and 9 to grayscale
```

The above syntax is defined below.

```
PEN(#)  –  where # defines a numeric pen value
(#,#,#)  –  where # defines the color table number to modify
/rgb  –    defines the output color in RGB values
```

PEN TABLE DEFINITIONS

Use the Pen Table Definitions section of the driver file to define the specific pen table to be used during the plot process.

```
pentable=SITE:pentables\XYZ_Color.tbl
```

The syntax used is defined below.

SITE: –   where # defines a numeric pen value

(#.#.#) –   where # defines the color table number to modify

/rgb –   defines the output color in RGB values

PAGE SETUPS

Use the Page Setups section to modify the delivered paper sizes in the driver file. The last size listed is the default paper size in the Print dialog.

```
size=(8.5,11.0)/num=0/off=(0.0,0.0)/name=Asize
size=(11.0,17.0)/num=0/off=(0.0,0.0)/name=Bsize
size=(17.0,22.0)/num=0/off=(0.0,0.0)/name=Csize
size=(24.0,36.0)/num=0/off=(0.0,0.0)/name=Dsize
size=(42.0,30.0)/num=0/off=(0.0,0.0)/name=Fsize
size=(48.0,36.0)/num=0/off=(0.0,0.0)/name=Esize
```

LINE STYLE DEFINITIONS

Use the Line Style Definitions section to define line styles 1 through 7. You can modify the plot appearance of these "on-screen" line styles using the driver file. The following is a sample of the delivered line styles.

```
style(1) = (14,42)/nohardware              ; style = dot
style(2) = (70,42)/nohardware              ; style = med dash
style(3) = (168,56)/nohardware             ; style = long dash
style(4) = (112,42,28,42)/nohardware       ; style = dot-dash
style(5) = (56,56)/nohardware              ; style = short dash
style(6) = (84,28,28,28,28,28)/nohardware  ; style = dash-dot-dot
style(7) = (112,28,56,28)/nohardware       ; style = long-dash short-dash
```

Line Style 0 ————————————————

Line Style 1 --------------------------------

Line Style 2 - - - - - - - - - - - - - - - - -

Line Style 3 – – – – – – – – – – – – – –

Line Style 4 -·—·—·—·—·—·—·—·—

Line Style 5 ————————————————

Line Style 6 —·—·—·—·—·—·—·—·—

Line Style 7 ————————————————

You can modify any of these line styles for your corporate requirements. For example, modify line style 4 to more closely resemble a center line style. The following is a sample of a modified line style 4.

```
style(4) = (1000,50,75,50)/nohardware   ; style = CENTER LINES
```

Line Style 4 ————— · ————— · —————

LINE WEIGHT DEFINITIONS

Use the Line Weight Definitions section to define line weights 0 through 32. You can modify the plot appearance of these "on-screen" line weights using the plot driver file. You can use mm values to define the line weights in MicroStation and to match line weight in AutoCAD's plot style tables. The following is a sample of the delivered line weights.

```
weight_strokes(mm)=(0.25, 0.50, 0.75, 1.00, 1.25, 1.50, 1.75, 2.00,
2.25, 2.50, 2.75, 3.00, 3.25, 3.50, 3.75, 4.00, 4.25, 4.50, 4.75, 5.00,
5.25, 5.50, 5.75, 6.00, 6.25, 6.50, 6.75, 7.00, 7.25, 7.50, 7.75, 8.00)
```

The following is a sample of modified line weights.

```
weight_strokes(mm)=(.09,.18,.25,.30,.35,.40,.50,.60,.70,1.00,1.10,1.20
,1.30,1.40,1.50,1.60,1.70,1.80,1.90,2.00,2.10,2.20,2.30,2.40,2.50,2.60
,2.70,2.80,2.90,3.00,3.10)
```

USING WINDOWS PRINTER DRIVERS

The majority of the driver file settings discussed are identical.

Page Setups

Use the Page Setups section to automate the paper size selections available in the driver file. This setting can override any default Windows printer preferences.

```
sysprinter /name="HP 750C"          /form=Arch E1 30 x 42 in
```

Page Layout

Use the Page Layout section to automate the paper layout and orientation. This setting can override any default Windows printer preferences.

```
sysprinter /name="HP 750C"          /form=Arch E1 30 x 42 in
                                     /orientation=landscape
```

Plot Options

Use the Plot Options section to automate the default plot options available for the Windows printer driver.

```
;border /pen=1 /time /filename /text_height=0.4
;fence_outline/pen=1
```

You can change these settings manually during the plot process.

NODISPLAY: Fence Boundary

Plot Border

USING SHEETS TO PRINT

MicroStation V8 provides a "paperspace-like" environment for layouts and printing. You can import and export both modelspace and paperspace environments while maintaining all functionality between the two applications. However, most MicroStation users do not use sheets in the same manner as AutoCAD. This is a new environment for MicroStation, and there has been some resistance to this environment. However, as an AutoCAD user you can understand better than anyone; after all, many AutoCAD users resisted, and still resist, the paperspace environment.

I encourage you to remember back to your first explorations into paperspace; they weren't always enjoyable experiences either. Use this background knowledge and understanding of paperspace to encourage and foster more widespread use of this environment within the MicroStation community.

There are specific rules that must be adhered to for the sheet and paperspace environment to be interchangeable. Only sheets defined using the 1:1 Sheet technique are exactly interchangeable with AutoCAD. The Plot Scale Sheets are not defined with exact AutoCAD functionality.

Sheets

Sheets can be defined in real-world units at 1:1, or at a plot scale (e.g., 1" = 50').

1:1 SHEETS
Sheets defined using this method are established in real-world units and the paper size is 1:1. The design model is then referenced into the sheet at the preferred plot scale for each discipline-specific drawing. These reference files are then clipped as needed to fit into the 1:1 sheet layout. These clipping boundaries will become the viewport boundaries when transferred into the AutoCAD application.

PLOT SCALE SHEETS
Sheets defined using this method are established in plot-scale units and the paper size is exaggerated to fit around a true scale design. The design model is then referenced into the sheet at a scale of 1:1. These reference files are then clipped as needed to fit into the sheet layout. These clipping boundaries

will become the viewport boundaries when transferred into the AutoCAD application.

AutoCAD Command Comparison

AutoCAD		MicroStation
Mouse pick: Layout1 Layouts	Keyboard: *LAYOUT*	Use a sheet model to simulate the behavior and usage found in AutoCAD's layouts. You can access a specific sheet using the View Groups toolbar to navigate between design models and sheet models.
Scaled Viewport		Use scaled references into the sheet model to simulate the behavior of a scaled viewport into paperspace. The edge of the clip boundary defines the edge of the viewport. The scale of the reference file determines the scale of the viewport.

In Exercise 9-1, following, you have the opportunity to practice printing sheets.

EXERCISE 9-1: PRINTING WITH A SHEET

In this exercise you will learn to use basic printing techniques available in MicroStation. Use files, views, or fences to define plot areas and create flexible printed output.

1   Open the design file *PLOT_SHEET1.DGN*.

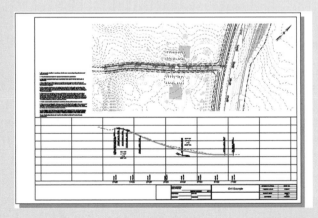

In this example, the sheet is already defined as 34 x 22, D size, using the annotation scale of 1" = 50'.

2 Select **File** > **Print** to access the Print dialog

Note that the following settings are automatically set from values defined in the sheet.

Area: Sheet
View: View 1
Paper: ANSI D

Sheet Properties

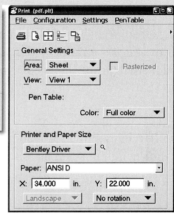

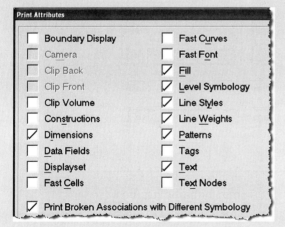

3 Use the Show Preview button to expand the Print dialog to preview the print.

4 Set the printer to use a Bentley driver and select the *PDF.PLT* plot driver file using the Browse button. **Q**

5 Select the Preview button to open the larger preview window.

Close this window after previewing the current print.

6 Select the Print Attributes button and verify the settings in the figure above left.

7 Activate the Print Border option and key in the comment *Exercise 9-1* to label this print.

Click on OK to save the settings changes.

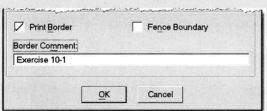

8 Select the Print button to send the print to a PDF file and save it to the following location.

   C:\MST_for_ACAD\out\Exercise 9-1.PDF

9 Open the PDF file to check the printed output.

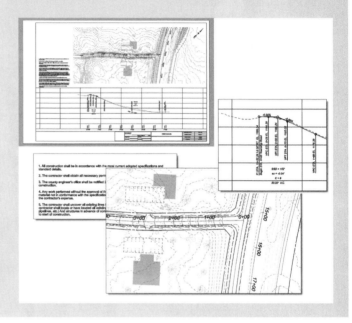

In Exercise 9-2, following, you have the opportunity to practice using a fence for printing.

EXERCISE 9-2: PRINTING WITH A FENCE

Next we want to print using the design model and a fence to define the plot area.

1 Open the design file *PLOT_DESIGN1.DGN*.

2 Select the Place Fence tool and snap to P1 and P2 to place the fence using the dotted shape representing the edge of the paper border.

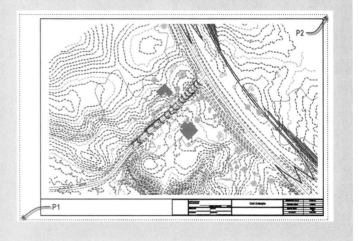

Be sure to snap to the dotted outline to automate the plot dialog settings accurately.

3 Select **File** > **Print** to open the Print dialog.

Note that the following settings are automatically set from the fence definition.

Area: Fence

4 Set the Paper size to ANSI D and select the Maximize Print Size button to refresh the preview and plot size.

The X and Y paper size should read 34.000 x 22.000 and the scale should read 50.00 automatically.

5 Set the printer to use a Bentley driver and select the *PDF.PLT* plot driver file.

6 Select the Print Attributes button and verify the settings shown at right.

7 Activate the Print Border option and key in the comment *Exercise 9-2* to label this print, as shown below.

Click on OK to save the settings changes.

Print Attributes

☐ Boundary Display	☐ Fast Curves
☐ Camera	☐ Fast Font
☐ Clip Back	☑ Fill
☐ Clip Front	☑ Level Symbology
☐ Clip Volume	☑ Line Styles
☐ Constructions	☑ Line Weights
☑ Dimensions	☑ Patterns
☐ Data Fields	☐ Tags
☐ Displayset	☑ Text
☐ Fast Cells	☐ Text Nodes

☑ Print Broken Associations with Different Symbology

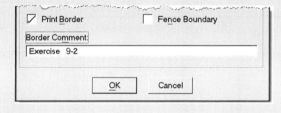

8 Select the Print button to send the print to a PDF file and save it to the following location.

C:\MST_for_ACAD\out\Exercise 9-2.PDF

9 Open the PDF file to check the printed output.

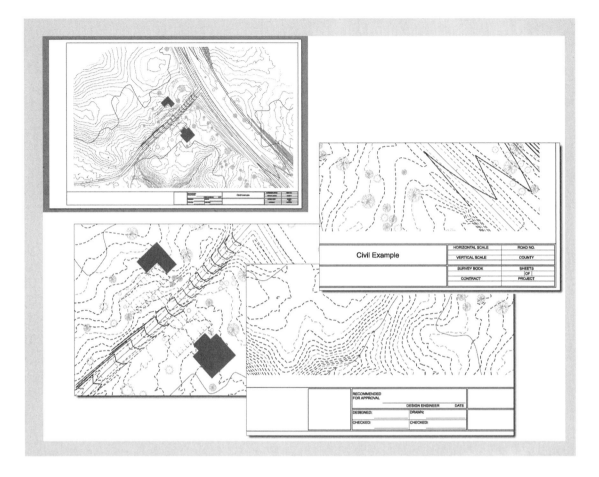

In Exercise 9-3, following, you have the opportunity to practice using pen tables.

EXERCISE 9-3: USING PEN TABLES

In this exercise you will learn to use pen tables to modify the printed output without having to make any changes to the design file.

1   Open the design file *PEN_TABLES1.DGN*.

First, let's learn to correct the text display problems where some of the text is below the filled building shapes. We can accomplish this by bringing all elements to the front and then moving the building shapes to the back during the print process.

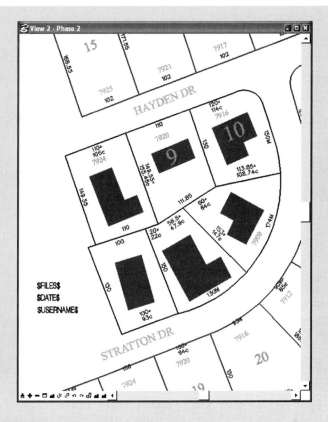

2   Select **File > Print** to open the Print dialog.

3   From the Print dialog, select the pull-down menu **PenTable > New** and create a new pen table named *EXERCISE 9-3.TBL*.

4   Select **Insert > Insert New Section Above** and name the section *Everything_to_Front*.

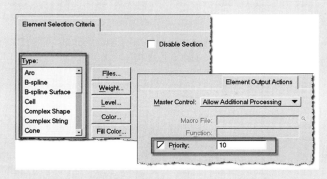

5   From the Element Selection Criteria tab, select all element types in the list. Select the Level button and select all levels except *Buildings*.

Pick OK to save these changes.

6   From the Element Output Actions tab, activate the Priority setting and set all elements to a priority of 10.

7 Select **File > Save** to save the pen table.

8 Select **Insert > Insert New Section Below** and name the section *Buildings_to_Back*.

9 From the Element Selection Criteria tab, select the element type *SHAPE* in the list.

10 Select the Level button and select the level *BUILDINGS* to search for shapes on this level only.

Click on OK to save these changes.

11 From the Element Output Actions tab, activate the Priority setting and set all building shapes to a priority of 1.

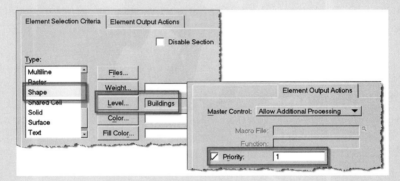

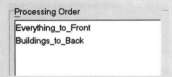

12 Highlight the *NEW* section in the Processing Order list and then go to the pull-down menu **Edit > Delete Section** to remove this section from the pen table.

13 Select **File > Save** to save the pen table.

14 Select the Windows Close button to close the Modify Pen Table dialog.

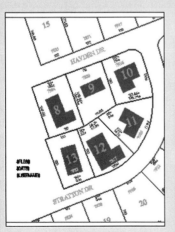

15 Use the Refresh Preview button to refresh the print preview and to verify that your pen table is working and that all text is above the filled building shapes.

Next, let's add a fill to the lot shapes.

16 From the Print dialog, select **PenTable > Edit** to add parameters to the pen table *EXERCISE 9-3.TBL*.

17 Select **Insert > Insert New Section Below** and name the section *FILL_LOTS*.

18 From the Element Selection Criteria tab, select the element type *COMPLEX SHAPE* in the list.

19 Select the Level button and select the level *LOT_SHAPES* to search for shapes on this level only.

Click on the OK button to save these changes.

20 In the Element Output Actions tab, establish the following settings.

    Fill: ON

    Fill Color: By RGB (and select a green color)

Processing Order

Fill_Lots
Everything_to_Front
Buildings_to_Back

21 Use the Up button to move the *Fill_Lots* section to the top of the Processing Order list so that the filled lots will process first.

22 Select **File > Save** to save the pen table.

23 Select the Windows Close button to close the Modify Pen Table dialog.

24 Use the Refresh Preview button to refresh the print preview and to verify that your pen table is working and that all lots are filled with the selected green color.

Next, let's use text substitution to populate "intelligent" text automatically at plot time. Text substitution will replace the existing text characters with the current file name, current date, and current user name.

    $FILES$
    $DATE$
    $USERNAME$

25 From the Print dialog, select **PenTable** > **Edit** to add parameters to the pen table *EXERCISE 9-3.TBL*.

26 Select the Text Substitution button to assign the text substitutions needed.

<u>T</u>ext substitutions...

27 Select **Edit** > **Insert Design File** > **Short**.

Note the default text format for the ACTUAL text placed in the design file. The ACTUAL text *$FILES$* is substituted with the pen-table-driven value *_FILES_*, which will print as a short file name with no path.

You can change the value of the ACTUAL text in the design file here. The use of the $ character is intentional because this character is rarely found in standard engineering drawing text.

28 Select **Edit** > **Insert Date**.

29 Select **Edit** > **Insert New** to create a custom text substitution.

Key in the following values for the user name text substitution.

> *Actual:* $USERNAME$
> *Replacement:* $(username)

HINT: $(username) *is the Windows syntax for retrieving the name of the user currently logged in to Windows.*

30 Select the Windows Close button to close the Text Substitution dialog.

31 Select **File** > **Save** to save the pen table.

32 Select the Windows Close button to close the Modify Pen Table dialog.

33 Use the Refresh Preview button to refresh the print preview and to verify that your pen table is working and that all text has been substituted with current text values.

Batch Printing

The need to print multiple drawings and sets of drawings is common in most organizations, and MicroStation provides very efficient and practical tools for processing these drawings unattended. You can create a *.JOB* file with preferred batch print settings and use these settings throughout the project life cycle. This *.JOB* file will contain all printer, print area, layout, and display settings, as well as a list of drawings to be printed.

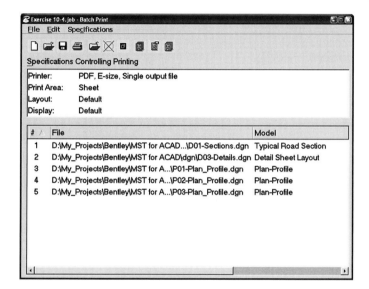

PRINTER

The Printer setting defines the print driver file to be used. This can be a Bentley driver or Windows driver file. This driver file will control the additional options available when using the Batch Print function.

The delivered options are:

- ❑ Default
- ❑ HPDesignJet, E-size, Landscape
- ❑ HP-GL/2, D-size, Landscape
- ❑ PDF, E-size, Single output file
- ❑ PostScript, Color, Landscape

For example, *PDF.PLT* will allow output to files only, whereas *HP 750C.PLT* will allow output to plot files or hardware devices. The batch print output files are named the same as the design file, with varying extensions based on your preferences. By default, the extension will increment automatically from such as *filename.000*, to *filename.001*, to *filename.002*, and so on.

There are several substitution strings you can use to control the naming of batch output files.

%d   DGN file name

%b   Boundary counter; 3 digits

%e   Extension

%j   Job name

%p   Printer counter; 3 digits

%m   Model name

%x   DGN file extension

PRINT AREA

This setting defines the area of the drawing to be printed. The default choices follow. You must customize your drawing files for the "user should customize" options to function properly. The delivered options are:

❑ Default

❑ Sample Plot Cell (user should customize)

❑ Sample Plot Shape (user should customize)

❑ Sample Saved View (user should customize)

❑ Sheet

You can use the view window to define your print area with the options Fit, Fit Master, Fit All, and others.

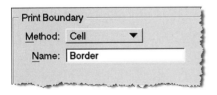

Plot Cell

Using this setting requires you to specify a cell name whose extents will be used as the print area.

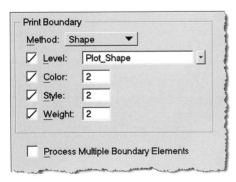

Plot Shape

Using this setting requires you to define the shape attributes whose extents will be used as the print area.

LAYOUT

This setting defines how Batch Print determines the location and size of the area to be printed. The following size options are available.

❑ Default

❑ 1" = 1 master unit

❑ 1/4" = 1 master unit

You can use the following parameters to define the layout for batch printing.

❑ Maximize

❑ Scale

❑ % of minimum size

❑ X size

❑ Y size

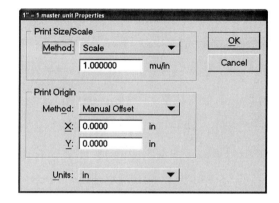

DISPLAY

The Display setting defines the appearance of printed elements. This is the batch printing equivalent of the View Attributes controls. This is the place to control the printed output regardless of the design file's saved settings. For example, use these settings to permanently deactivate the plotting of enter data fields, text nodes, or broken associations. Use it to permanently activate the plotting of items such as line weights, line styles, or text regardless of view attribute settings.

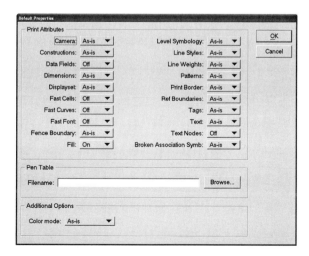

In Exercise 9-4, following, you have the opportunity to practice batch printing.

EXERCISE 9-4: USING BATCH PRINT

In this exercise you will learn to use the Batch Print utility to create multiple prints easily using standard settings and controls.

1   Open the design file *BATCH_PRINT1.DGN*.

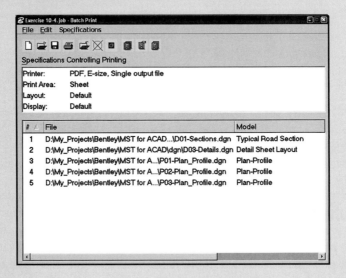

2   Select **File > Batch Print** to open the Batch Print dialog.

3   Highlight the Printer specification from the list and click on the Select Specifications button (the green button).

Select Printer PDF, E-size, and Landscape, and then click on OK to close the dialog.

4   Highlight the Print Area specification from the list and then click on the Select Specifications button.

Select the print area Sheet and then click on OK to close the dialog.

5   Leave the Layout and Display specifications set to default for this exercise.

6 Select **File** > **Save** to create a file of batch print specifications and then save it to the following location.

C:\MST_for_ACAD\out\Exercise 9-4.JOB

7 Click on the Add Design Files button to add the design files you want to print.

Select the following files to be printed.

- ❏ *D01-Sections.dgn*
- ❏ *D03-Details.dgn*
- ❏ *P01-Plan_Profile.dgn*
- ❏ *P02-Plan_Profile.dgn*
- ❏ *P03-Plan_Profile.dgn*

8 Remove all DEFAULT models from the drawing list.

9 Save these changes to the *.JOB* file.

10 Click on the Print button and print all files as PDFs to the following location¨

C:\MST_for_ACAD\out

11 Open the PDF file to check the printed output.

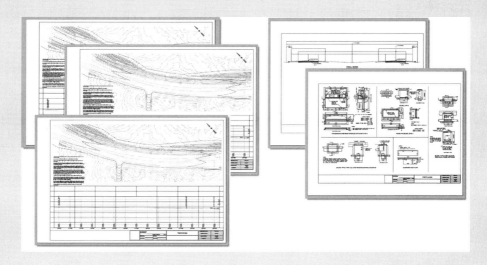

10: Working with DWG Files

Chapter Objectives:

❑ Learn what aspects of the DWG are compatible with MicroStation

❑ Learn how to make MicroStation more or less DWG compatible

❑ Learn to batch convert files between the DGN and DWG file formats

❑ Learn how to remap data from the DGN to the DWG file format

Because of the many changes to the DGN file format, working with DWG drawing files is easier than ever before. The files are not identical and there are still some aspects to consider when exchanging information, but this process has been greatly improved.

Rules for Success

There are some rules it is wise to follow when working with data that must be exchanged between MicroStation and AutoCAD.

❑ Enforce strict CAD standards for multi-CAD projects.

❑ Do not convert data that does not need to be converted.

❑ Do not round-trip converted data more than once.

❑ Know the limitations of each CAD program and work within them.

❑ If you have to deliver a certain format, begin with that format's seed file.

With that said, we all know that sometimes these things are out of your control, and you have to work with what you get.

WHY STRICT CAD STANDARDS?

The enforcement of CAD standards is always important on every project, but when you are working with consultants and outside vendors using different CAD applications, the quality and predictability of the data is critical. Nothing can destroy the success of this environment quicker than unpredictable data. You must know the specifics of the data in order to control the exchange. The following items deserve close attention.

- ❑ Fonts
- ❑ Level names
- ❑ Symbol cells or blocks

- ❑ Line styles
- ❑ Line weights
- ❑ ByLevel/ByCell attributes

Fonts

You can really use any type of font you prefer, as long as it is a "common" font to one of the CAD applications, or a delivered Windows font. Do not use custom fonts unless you are prepared to manage the conversion. Most fonts exchange well and the conversion process is quite simple using the Remapping and Batch Convert utilities.

Both applications support the use of TrueType fonts, and thus it might simplify things to use standard Windows-delivered TTF fonts where possible. They are easy to read on the screen, they plot well, and they do not require the use of line weight. A second suggestion is to use AutoCAD SHX fonts, because both applications can read them, or you can use MicroStation fonts and manage their conversion using the DWG remapping options.

One important side note on the font conversions performed from MicroStation is that when you save a DGN file to the DWG format the default setting will convert all used MicroStation fonts to SHX fonts automatically. Initially, you might think that this is good. However, it may cause problems for the AutoCAD user. If you open the DWG using AutoCAD it will look fine because you have these converted fonts on your system. When you send this file to someone else, however, they will not have these fonts on their system and the file will not look the same. You have two solutions to this problem.

1   Send the converted MicroStation SHX fonts with your DWG files.

2   Convert the MicroStation fonts to a delivered AutoCAD font using a remapping file.

Refer to the "Recommended DWG Settings" section for further information on these settings.

Line Styles

MicroStation and AutoCAD both read line styles differently, and thus you should manage how this conversion takes place. By default, MicroStation will not drop unsupported line styles. Thus, if unexpected line styles are encountered they may appear as solid lines in AutoCAD. You have two solutions to this issue.

1 Manage the line style conversions using a remapping file.

2 Drop all unsupported line styles during the conversion. This is the easiest solution but not the most user-friendly.

Level Names

MicroStation has used level names for several years, but legacy users were so familiar with level numbers that they were not widely adopted. In the V8 file format, the use of level names became mandatory, although numbers do still exist. Fortunately, most users are making this transition and learning to embrace the level name capabilities. If level names are not defined, V8 will rename the numbers per the convention *Level #*. The use of a remapping file can easily convert these very unfriendly level names to more appropriate level names.

ByLevel / ByCell

One of the many new concepts commonly misunderstood is ByLevel attributes and their benefit in day-to-day use. This concept has been in AutoCAD for many years and is a very efficient way of managing levels. It is not required, but it deserves strong consideration when defining V8 CAD standards. This is especially important for those exchanging data with AutoCAD, because this is how AutoCAD utilizes ByLayer symbology. The term *ByLevel* defines the "control" of element attributes such color, line weight, and line style. Using the ByLevel functionality allows you to assign element attributes to the actual level, eliminating the need to set color, line weight, and line style independently.

The ByCell functionality allows for a "tweak factor" to the ByLevel attributes. Using this setting for an attribute value provides you with the ability to modify a color, line weight, or line style "on the fly" when needed, but still automatically uses ByLevel by default. If you understand ByLayer/ByBlock in AutoCAD, you understand ByLevel/ByCell in MicroStation. The ByCell setting is not activated by default. If you need this setting activated, you must modify the workmode capability settings.

Symbols

The use of symbols in generating engineering drawings has been used since the days of board drafting, and symbols continue to be useful in CAD drafting today. MicroStation uses cells as symbols, and in the V8 file format they are very compatible with DWG blocks. Cells and blocks can be exchanged easily between the CAD applications, and (depending on the use of ByLevel/ByLayer and ByCell/ByBlock) can be exchanged transparently.

Why Not Convert?

One of the most important rules to remember is to never convert DGN to DWG, or DWG to DGN, if you do not have to. MicroStation V8 allows you to reference DWG files directly. If you are being provided DWG data and you plan to use it as reference data there is no need to convert it to DGN. V8 also allows you to open the DWG directly and make edits without the need to convert it to DGN. Leave data in its original format whenever possible to maintain data integrity.

Why Not Round-trip?

Even if you follow the previous rule most of the time, you know there will be times when the data has to be converted. In this event, the next rule to remember is to minimize the trips the data has to take. To round-trip data, you are taking the data from its original format to a "foreign" format, and back again. This round trip can cause unpredictable results. You should convert the data one-way only, and avoid a return trip whenever possible. Every time you round-trip data you increase the chance of inconsistencies creeping into the data.

DGN AND DWG WORKMODES

What Are Workmodes?

Workmodes are "environments" available in MicroStation V8 that control what capabilities are available and how various file types are processed.

There are basically three different workmodes available in V8.

❑ DGN workmode

❑ DWG workmode

❑ V7 workmode

DGN WORKMODE

The DGN workmode provides you with a fully functional DGN file with all V8 features and enhancements. This is the default workmode when a V8 DGN file is opened.

DWG WORKMODE

The DWG workmode restricts the functionality in V8 DGN files to maintain compatibility with DWG files. This is the default workmode when a DWG file is opened.

V7 WORKMODE

The V7 workmode restricts the functionality in DGN files to maintain compatibility with V7 DGN files. This workmode can only be activated by defining the following configuration variable.

    MS_OPENV7=3

Active Workmode

You can determine the active workmode from the status bar in MicroStation. Click on the Workmode icon to check the active workmode settings.

DWG workmode   DGN workmode   V7 workmode

DWG Workmode Limitations

This workmode is considered somewhat restrictive. However, this allows the file to be 100% compatible with the AutoCAD DWG file format. Keep in mind that you can "tweak" these settings as needed to allow for more flexibility while working in DWG files. Below is a list of restrictions when using the DWG workmode.

Annotation: Flags are disabled.

Area Patterning: Area patterns are replaced with AutoCAD hatching. The Pattern settings window lists the patterns in the DWG pattern file *areapat.pat*. This file contains a DWG pattern definition for each pattern cell in the supplied cell library *areapat.cel*.

Cells: Only shared cells are allowed, and new cells cannot be created in an attached DWG file. New cells must be created in a DGN or CEL library file.

Colors: AutoCAD does not allow for custom color tables.

Complex Elements: Spline curves are not allowed as part of a complex shape or complex chain.

Curves: B-splines are not supported.

Design History: Design History is disabled.

Dimension Driven Design: Dimension-driven tools are disabled.

Dimensions: Various dimension settings are disabled.

Element Class: Active class is set to primary, and construction class cannot be activated.

Grid Orientation: Grid alignment with view is not available.

Raster Files: Intergraph, SUN, georeferenced TIFF, and LMG files cannot be imported.

Line Styles: Line styles 1 through 7 are disabled, and only custom line styles are allowed.

Models: Restricted to a single design model and multiple sheet models.

References: Self-referencing and clip masking are disabled.

Saved Views: Various limitations on saved views and sheet models.

Symbology: Element symbology is disabled and ByLevel symbology is preferred.

Tags: Tags are attached only to shared cells.

Text Fonts: SHX and TTF fonts are supported.

Text Styles: Various text parameters are disabled.

View Groups: View groups cannot be created, modified, or deleted.

View Windows: The black-to-white background user preference is disabled. Background color can be modified using the DWG Open Options feature.

V7 Workmode Limitations

There are some limitations when you use the old V7 DGN file format.

View Groups: View group creation is disabled.

Levels: Level creation is disabled.

Models: All Model creation is disabled.

References: 3D-to-2D reference file attachment is disabled. Reference associations are disabled. References to models in V8 are disabled. References to DWG files are disabled.

Modifying the Workmode Capabilities

Many of the workmode restrictions can be modified using the configuration settings found in the following workmode configuration file.

C:\Program Files\Bentley\Program\MicroStation\config\system\workmode.cfg

Do not edit the delivered *workmode.cfg* file. Keep the original unchanged because new versions will overwrite your changes. Use this file to investigate what changes you can make and place your workmode configuration changes in a site-specific or standards configuration file within your workspace environment.

Workmode Configuration File Syntax

A workmode capability can be enabled or disabled using the following syntax.

```
<WORKMODE VARIABLE> <Operator> <Prefix> <CAPABILITY_NAME>
```

The components of this syntax are as follows.

WORKMODE VARIABLE: Sets the workmode for the file.

OPERATOR: Appends (>) or prepends (<) the capability variable value to the workmode variable. Do not use the (=) sign for these definitions.

PREFIX: Indicates whether the function is to be enabled (+) or disabled (-).

CAPABILITY_NAME: The name of the function to be enabled or disabled.

WORKMODE VARIABLES

The following workmode variables can be set.

_USTN_CAPABILITY: The capability applies in all workmodes.

_USTN_CAPABILITY_DGN: The capability applies in V8 workmode only.

_USTN_CAPABILITY_DWG: The capability applies only in DWG workmode.

_USTN_CAPABILITY_V7: The capability applies only in V7 workmode.

OPERATORS

To define configuration variables, you can use the following operators.

< Prepend the capability variable value to the workmode variable settings

> Append the capability variable value to the workmode variable settings

PREFIXES

Use the following prefixes to activate or deactivate capability variables.

+   Activate the capability variable

-   Deactivate the capability variable

The following are examples of capability definitions.

CAPABILITY_BYCELL: If enabled, allows ByCell settings for element attributes. The following, for example, adds the ByCell capability in *all* workmodes:

_USTN_CAPABILITY < +CAPABILITY_BYCELL

CAPABILITY_LEVELS_GLOBALDISPLAY: If enabled, allows levels to be turned on and off using global display. The following, for example, adds the capability to create levels in DWG workmode:

_USTN_CAPABILITY_DWG < -CAPABILITY_LEVELS_GLOBALDISPLAY -

CAPABILITY_LEVELS_CREATE: If enabled, controls levels the ability to create new levels in a file. The following, for example, removes the capability to create levels in DWG workmode:

_USTN_CAPABILITY_DWG < -CAPABILITY_LEVELS_CREATE -

RECOMMENDED DWG SETTINGS

The following sections discuss recommendations for the Open DWG and Save As DWG options.

Open DWG

The default settings for opening a DWG file are set exactly as needed, with no changes unless you have a specific problem to fix.

Save As DWG

The default settings for saving a DWG file are primarily set as needed. However, the changes discussed in the following sections are recommended.

BASIC UNITS
You should modify the Units setting to Sub Units if you work with a base unit of inches.

```
□ Basic
       DWG Version:                              2000/2000i/2002
       Level Display:                            Global
       Units:                                    Sub Units
       Line Code Scale (Design Units/Cycle):     0.0000
       Use Level Symbology Overrides             □
       Preserve MicroStation Settings            □
       DWG Seed File:                            seed.dwg
```

REFERENCES
If you use relative reference paths, you should make the following change to the References setting.

```
□ References
       Create Overlays for References (No Live Nesting)    □
       Save Path:                                Relative to Master File
       Map Logical Names to XRef Block Names      ☑
```

CELLS
Make the following changes if you do not use ByLevel attribute settings for your MicroStation cell color, line weight, and line style.

```
□ Cells
         Create Single Block For Duplicated Cells     ☑
         Create Scaled Blocks                          ☑
         □ Create Block Entities With "By Block" Properties
              Level (Layer 0)                          ☑
              Color                                    ☑
              Line Style                               ☑
              Line Weight                              ☑
```

FONTS

Modify the name setting for Text Styles to not include the *STYLE-* variable so that DWG file style names will not obviously be from MicroStation and to match the AutoCAD style names more closely.

⊟Fonts	
Code Page for Dwg File	English - 1252
Text Style Name Template	%s
Convert MicroStation fonts to AutoCAD fonts	
SHX Output Directory:	C:\DWGfonts\

You should control the folder where the MicroStation SHX converted files are stored so that you have easy access to them for submittals and so that they do not end up in your default AutoCAD installation. You will have a better test environment if you leave your AutoCAD installation untouched by MicroStation.

The remaining settings are adequate for the majority of DWG output, but remember to revisit this section if you encounter problems with specific files.

STANDARDS FOR THE DWG SAVE AS OPTIONS

Using MicroStation V8 is very robust for converting DGN data to DWG, or DWG data to DGN. You can easily convert files individually or in bulk using batch process utilities. Standard conversion settings can be saved in resource files and remapping files to be used repeatedly, saving time throughout a corporate or project life cycle.

With the significant number of settings that can be defined, some discussed earlier in this chapter, the storage of your preferred settings is critical to your sanity. The good news is that you can do just about anything to the data. The bad news is that it can get a little confusing trying to organize and define all of the settings. When you complete these settings changes, MicroStation stores them in a resource file. This resource file can be a single resource file or several, depending on your projects and standards environments. Use the workspace to point to this resource file, and if necessary lock it so that critical settings cannot be changed haphazardly. Remember to test these settings early on to avoid problems later on.

The DWG Settings Resource File

So where is this resource file? By default it is located in:

C:\Program Files\Bentley\Home\prefs\dwgdata.rsc.

Whenever you make a change to the DWG options dialog, those changes are written to this resource file.

TIP: *If you want to revert back to the default DWG options resource settings, delete this file and MicroStation will create a new default settings resource file the next time you access the DWG options.*

What Is a Remapping File

A remapping file is a spreadsheet provided with MicroStation that contains specific parameters for handling and processing data. This data can be converted in many different directions, such as DGN to DGN, DGN to DWG, or DWG to DGN. So what parameters can you modify during this process? The following are categories provided for parameter mapping.

❑ Levels   ❑ Font

❑ Line Styles   ❑ Color

❑ Weight   ❑ Symbol

TIP: *Each sheet in the remapping Excel file contains instructions at the bottom of the page. Scroll down to find additional instructions for each sheet type. There is also a General Instruction sheet provided.*

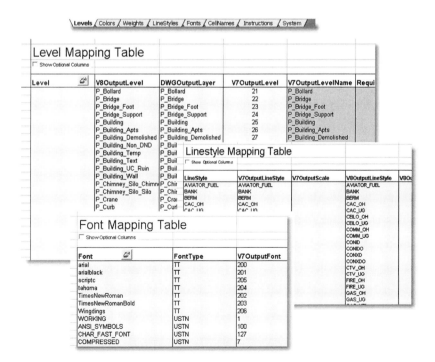

LEVEL MAPPING

Level mapping allows you to control how level data is distributed between level and layer structures, and to manipulate level values and evaluate and filter nonstandard levels into standard levels. The following parameters can be manipulated.

- ❑ Level name
- ❑ DWG output layer
- ❑ V7 output level name
- ❑ Level description
- ❑ Override color
- ❑ Override line style
- ❑ Override line style origin width
- ❑ ByLevel color
- ❑ ByLevel line style
- ❑ ByLevel line style origin width
- ❑ Global display
- ❑ Plot

- ❑ V8 output level
- ❑ V7 output level
- ❑ Required levels
- ❑ Level number
- ❑ Override line weight
- ❑ Override line style scale
- ❑ Override line style end width
- ❑ ByLevel line weight
- ❑ ByLevel line style scale
- ❑ ByLevel line style end width
- ❑ Element access

FONT MAPPING

Font mapping allows you to control how font data is converted between MicroStation, AutoCAD, and Windows, and to manipulate, evaluate, and filter font values from nonstandard fonts into standard fonts. The following parameters can be manipulated.

- ❑ Font name
- ❑ V7 output font
- ❑ DWG output font
- ❑ Output width factor

- ❑ Font type
- ❑ V8 output font
- ❑ DWG output font type
- ❑ Output height factor

LINE STYLE MAPPING

Line style mapping allows you to control how line styles are converted between MicroStation and AutoCAD, and to manipulate, evaluate, and filter line styles from nonstandard line styles into standard line styles. The following parameters can be manipulated.

- ❑ Line style name
- ❑ V7 output scale
- ❑ V8 output scale
- ❑ DWG output line style scale

- ❑ V7 output line style
- ❑ V8 output line style
- ❑ DWG output line style
- ❑ DWG output line style resource file

COLOR MAPPING

Color mapping allows you to control how color is utilized in both MicroStation and AutoCAD, and to manipulate, evaluate, and filter color values from nonstandard colors into standard colors. The following parameters can be manipulated.

- ❑ Color number
- ❑ V8 output color
- ❑ V7 output color
- ❑ DWG output color

WEIGHT MAPPING

Weight mapping allows you to control how line weight is exchanged between MicroStation and AutoCAD, and to manipulate, evaluate, and filter color values from nonstandard colors into standard colors. The following parameters can be manipulated.

- ❑ Weight numbers
- ❑ V8 output line weight
- ❑ DWG output color
- ❑ V7 output line weight
- ❑ DWG output line weight

SYMBOL MAPPING

Symbol mapping allows you to control how symbols are converted between MicroStation V7 and V8, and to manipulate, evaluate, and filter symbols from nonstandard cells into standard cells and blocks. The following parameters can be manipulated.

- ❑ Symbol name
- ❑ V7 output cell name

In Exercise 10-1, following, you have the opportunity to practice converting DGN files to DWG files.

EXERCISE 10-1: CONVERTING DGN TO DWG FILES

In this exercise you will learn how to save a V7 DGN file to the DWG file format. Learn what settings to change (and why) to ensure quality DWG output.

First, let's discuss how this old DGN file was set up using an older version of MicroStation. In V7, most files did not have level names (only numbers associated with the levels). V7 files also did not use the *ByLevel* attribute symbology, causing these files to be not as AutoCAD friendly as they could be.

1　Open the design file *DWG_SAVE1.DGN*.

2　To access the Save As DWG tool, go to the pull-down menu **File** > **Save As** and set the Save As Type setting to AutoCAD Drawing Files .DWG.

Click on the Save button to save as a DWG file.

3　Review the DWG/DXF Units dialog that appears and select the applicable units.

Set the Units setting to FEET

Do not select the Do Not Display Again setting unless this is the unit setting for all DWG files from this point forward. If you do, you can go to **File** > **Open** and set Files of Type to DWG. Then click on the Options button and select **Advanced** > **Display Unit Alert** and activate this setting to restore the Units Alert dialogs.

Click on OK to continue the file conversion.

DWG/DXF Units

MicroStation V8 requires that the file units be accurately specified in order for "True" scaling to be calculated correctly when working with cells and reference files.It is not possible to infer the units for the DWG or DXF file: "D:\My_Projects\Bentley\MST for ACAD\dgn_COMPLETED\DWG_SAVE1.dwg" for the following reason:

This file has linear units set to Decimal (LUNITS = 2). It is not possible to infer the units for files with this setting. If you know that the Design Center Units settings (INSUNITS) is correctly set for this file, using this setting is recommended. The current setting for Design Center Units is "Meters".

Units:　Feet　　　　　▼

☐ Do not display again (Use this setting for all DWG/DXF files of this type)

OK

> **HINT:** *This dialog is asking you what the "base" unit should be in the DWG file. Remember, AutoCAD does not have working units, so MicroStation is trying to determine what discipline this file should be configured for. Because this is a civil discipline file you should select Feet for English units.*

As you can see, the file appears to convert correctly, and if you were not an AutoCAD user you would think everything is fine. Let's look at some of the possible problems for the AutoCAD user.

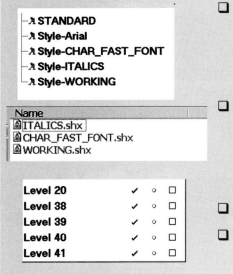

❑ Text styles were created with the *Style-* prefix, making it obvious that this file originated in MicroStation. This may be your preferred method. If so, ignore this problem. If not, you can change this so text styles have no prefix.

❑ The MicroStation fonts were converted to SHX files, requiring that these custom fonts be sent to the AutoCAD user. If this is your preferred method, leave as is. If not, use the remapping feature to remap these fonts to true AutoCAD-delivered fonts.

❑ The levels have numbered names, not logical names.

❑ All colors, line weights, and line styles are hard-coded to the individual elements.

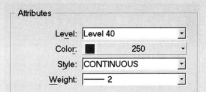

4   Reopen the design file *DWG_SAVE1.DGN*.

The first problem, text style names, is very easy to fix. Let's adjust that setting first.

5   Select the pull-down menu **File > Save As** and set the File as Type setting to DWG.

6   Click on the Options button to access the DWG Save As options.

7   Expand the Fonts portion of the General tab and remove *Style-* as the prefix for the Text Style Name template.

Leave the *%s* so that the text style will inherit the font name as the style name.

> **HINT:** *If you want all text to be converted to a single font, remove the %s and all text will be converted to a single text style.*

8   Deactivate the *Convert MicroStation fonts to AutoCAD fonts* setting because we do not want to use these custom fonts in our DWG file output.

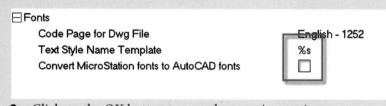

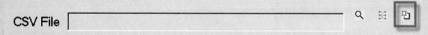

9   Click on the OK button to save these option settings.

10   Click on the Save button and overwrite any existing DWG file of the same name.

11   Select FEET for the base units and then click on OK to save the DWG file.

12   Select **Element** > **Text Styles** and verify the new text style names and their associated font settings.

The next few steps require a remapping file to direct MicroStation to perform the following remapping procedures.

❑   Remap MicroStation fonts to AutoCAD fonts

❑   Remap level numbers to useful layer names

❑   Remap all colors, line weights, and line styles to AutoCAD line styles

13   Reopen the design file *DWG_SAVE1.DGN*.

14   Select the pull-down menu **File** > **Save As** and set the File as Type setting to DWG.

15   Click on the Options button to access the DWG Save As options.

16   Select the Remap tab and then click on the Create CSV Remapping button.

Key in the name *DWG_SAVE1* for the remapping CSV file name.

Be patient at this point, because MicroStation is now launching the Microsoft Excel application with a specific spreadsheet you can use for these remapping features.

HINT: *You may get a security notification concerning the security levels in Excel allowing macros to run. You must allow this macro to run for this procedure to work properly.*

Remap the Fonts

17 In Excel spreadsheet, select the Fonts tab and fill in the following information to remap the fonts used.

Save the spreadsheet with these changes.

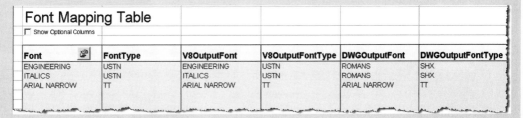

Font		FontType	V8OutputFont	V8OutputFontType	DWGOutputFont	DWGOutputFontType
ENGINEERING		USTN	ENGINEERING	USTN	ROMANS	SHX
ITALICS		USTN	ITALICS	USTN	ROMANS	SHX
ARIAL NARROW		TT	ARIAL NARROW	TT	ARIAL NARROW	TT

Font Mapping Table
Show Optional Columns

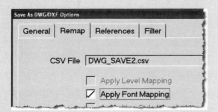

18 Reopen the file *DWG_SAVE1.DGN* and navigate back to the Remap dialog in MicroStation and activate the Font Mapping option.

Click on OK to save the remapping settings.

Save the DWG file and overwrite any existing files of the same name.

Use the Element Information command to verify the font for the lot numbers. The font should be converted to ROMANS.

Remap the Level Names

19 Navigate back to the Excel spreadsheet and select the Levels tab.

20 In Excel, click on the button to automatically populate the level names from the active design file.

This will add the DGN levels to the spreadsheet.

21 Fill in the following level names and descriptions in the spreadsheet and save the spreadsheet file.

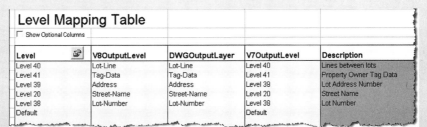

Level Mapping Table
Show Optional Columns

Level		V8OutputLevel	DWGOutputLayer	V7OutputLevel	Description
Level 40		Lot-Line	Lot-Line	Level 40	Lines between lots
Level 41		Tag-Data	Tag-Data	Level 41	Property Owner Tag Data
Level 39		Address	Address	Level 39	Lot Address Number
Level 20		Street-Name	Street-Name	Level 20	Street Name
Level 38		Lot-Number	Lot-Number	Level 38	Lot Number
Default				Default	

22 Scroll to the right of the spreadsheet to find the BYLEVEL columns for the following.

BYLEVELCOLOR   BYLEVELWEIGHT   BYLEVELSTYLE

Fill in the following information to remap the level attribute settings, and then save the spreadsheet file.

Save the spreadsheet with these changes.

Level Mapping Table

☐ Show Optional Columns

Level		V8OutputLevel	ByLevelColor	ByLevelWeight	ByLevelStyle
Level 40		Lot-Line	1	2	CENTER2
Level 41		Tag-Data	7	5	0
Level 39		Address	2	1	0
Level 20		Street-Name	6	1	0
Level 38		Lot-Number	5	1	0
Default			0	0	0

HINT: *Note that the BYLEVELSTYLE for the level Lot-Line is forced to the AutoCAD line type CENTER2. This allows you to modify the actual level settings during the conversion if needed. The AutoCAD line style CENTER2 must exist in the DGN file before MicroStation can use it during the conversion process. You can import them directly from the ACAD.LIN file using* **Custom Line Style > Edit > File > Import**.

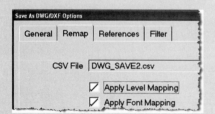

23 Reopen the file *DWG_SAVE1.DGN* and navigate to the Remap dialog in MicroStation and activate the Level Mapping option.

Click on OK to save the remapping settings.

Save the DWG file and overwrite any existing files of the same name.

24 Check the Level Manager in the DWG file for the new named levels with BYLEVEL symbology defined.

Remap the Elements

Next, we need to remap the element attributes to BYLEVEL symbology for color, line weight, and line style.

Color Mapping Table

☐ Show Optional Columns

Color	DWGOutputColor
%unmapped	%bylevel

25 Navigate back to the Excel spreadsheet and select the Color tab.

Fill in the following color information.

%unmapped will filter all element colors.

%bylevel will assign the BYLEVEL definition to all colors.

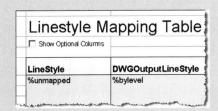

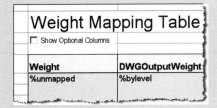

26 Select the LineStyles tab and fill in the following line style information.

%unmapped will filter all element line styles.

%bylevel will assign the BYLEVEL definition to all line styles.

27 Select the Weight tab and fill in the following line weight information.

%unmapped will filter all element line weights.

%bylevel will assign the BYLEVEL definition to all line weights.

28 Save these changes to the spreadsheet.

These remapping definitions will apply ByLevel symbology to all elements in the design file, which is identical to the ByLayer environment in AutoCAD.

TIP: *You might need to change the active linestylescale to see the new line styles in MicroStation. Use the following key-in as needed:*

`ACTIVE LINESTYLESCALE 18`

29 Navigate back to the Remap dialog in MicroStation and activate the Level Line Style and Weight Mapping options.

Click on OK to close the remapping dialog.

Save the DWG file and overwrite any existing files of the same name.

30 Use the Element Information command to check the attribute symbology of any graphical element.

Note that the color, line style, and line weight attributes are set to ByLevel.

Once the remapping file is completely defined with your CAD standards you can batch process your DGN-to-DWG conversions using the Batch Convert utility.

In Exercise 10-2, following, you have the opportunity to practice using the Batch Conversion utility.

EXERCISE 10-2: BATCH CONVERSION UTILITY

This exercise will batch convert several DGN files to the DWG format using a remapping file.

1   Open the design file *BLANK_V8.DGN*.

2   To access this utility, go to the pull-down menu **Utilities** > **Batch Converter**.

3   In the Batch Convert dialog, set the Default Output Format setting to *DWG* and the Default Destination setting to *C:\Training\MSTforACAD\DWGout*.

    If necessary, navigate to the correct folder using the Browse button.

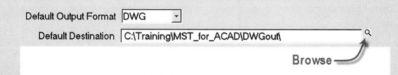

4   Click on the Add Files button or Directories button and add the following files to the conversion list.

<table>
<tr><td>

CB1.DGN

CB2.DGN

CB3.DGN

</td><td>

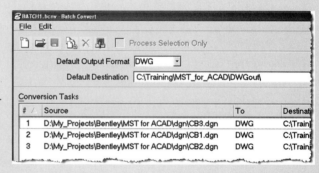

</td></tr>
</table>

5   Save this settings file to the following file location.

    C:\Training\MST_for_ACAD\ Resource\BATCH1.BCNV

6   To assign a remapping file to the batch conversion, go to the pull-down menu **Edit** > **DWG Save Options**.

7   Select the Remap tab and browse to the location of your remapping spreadsheet. For this exercise, we will use the following remapping spreadsheet.

    C:\Training\MST_for_ACAD\Resources\Remap_CBdrawings.xls

8   Click on the Process Batch Convert Job button to process all files in the list.

    Click on the Convert button to start the batch conversion.

9   Open the DWG files and verify the output.

11: Best Practices

Chapter Objectives:

❑ Determine project requirements and file format considerations.

❑ Explore DGN format considerations.

❑ Explore DWG format considerations.

The objective of this chapter is to assist you when working in a multi-CAD environment, which involves working with DGN and DWG data. The scenarios listed here are intended to help you get started down the right path and to help you avoid some of those stumbling blocks along the way.

Standards and Workflows

Using CAD standards and enforcing them can greatly improve the consistent flow of information between DGN and DWG files. The key to success is preparation and standardization. If you know what to expect in the data files, you can control the outcome. Unpredictable data will inevitably cause unpredictable results in both file formats.

The least desirable scenario is to round-trip files from one format to another and back again. Doing this can create data exchange inconsistencies, and the more trips through this process the data takes the more likely you are to see unexpected and undesirable results. Use the data files in their native format as as long as possible throughout the design process, and collaborate with outside resources (using their native formats whenever possible). You first need to determine which scenario most closely matches your production environment.

❑ Designing in DGN, using DWG resources, and delivering in DGN

❑ Designing in DWG, using DWG resources, and delivering in DWG

❑ Designing in DGN and delivering in DWG

Designing in DGN, Using DWG Resources, and Delivering in DGN

This scenario assumes that you will be creating, opening, and editing in the DGN file format and using DWG resource files for references but delivering the final output in the DGN format. In this type of project you would not need to restrict any of your MicroStation capabilities because you would be retaining the DGN file format through to the final output. This scenario may require some modifications to your CAD standards in order to manage the DWG references and their symbology. However, this can easily be handled using level symbology or pen tables. In fact, you might consider using the AutoCAD plot style tables to help you develop the MicroStation pen tables as needed.

Designing in DWG, Using DWG Resources, and Delivering in DWG

This scenario assumes that you will be creating, opening, editing, and delivering in the DWG file format. In this type of project you will be working in the DWG workmode, and some MicroStation capabilities will be disabled. This environment may be too restrictive and you can loosen the restrictions using the workmode capability settings. However, be sure to test any of these modifications in the AutoCAD environment to ensure complete compatibility.

Designing in DGN and Delivering in DWG

This scenario assumes that you will be creating, opening, and editing in the DGN file format and delivering DWG as final output. In this type of project you want to use MicroStation and all of its tools and features, but you have to deliver the final output in DWG file format. You could work in the DWG workmode, but this would restrict some of the MicroStation capabilities.

A better solution would be to work with DGN files for as long as possible before sending them to DWG. The key to success using this scenario is starting the project correctly. Use a controlled standards environment and use as many of the DWG-compatible features as possible. For example, consider the following.

❑ Use tag data in cells rather than enter data fields. Tag data will migrate to DWG as block attributes, which provide editable text in AutoCAD blocks.

❏ Use only one design model. AutoCAD does not allow for multiple modelspaces in a single DWG file. Multiple sheet models are acceptable because AutoCAD supports multiple paperspaces.

❏ Use text styles and dimension styles whenever possible.

❏ Use AutoCAD line types whenever possible. However, you can remap line styles during the Save As DWG process.

❏ Use no nesting or live nesting with reference file attachments. AutoCAD allows for "all or nothing" reference file attachments. No nesting will not inherit nested reference files. Live nesting will inherit all nested reference file attachments. Most AutoCAD users prefer no nesting when referencing other files.

> Overlay = No Nesting
> Attachment = Live Nesting

Use a logical number for the nest depth associated with live nesting. After all, do you really need to go 10 files deep? Probably not. Using a nest depth of 2 or 3 is usually sufficient. If in doubt, ask the DWG recipients if they prefer overlay or attached references.

❏ Set the DWG Open and DWG Save As settings carefully. These settings are stored in the following file, which can be placed on a server so that you can control all project team members' DWG settings.

> *C:\Program Files\Bentley\Home\prefs\dwgdata\dwgsettings.rsc*

The following DGN-only capabilities will not transfer to the DWG file format.

❏ Design history

❏ Custom line styles (require custom AutoCAD definitions)

❏ Custom patterns (require custom AutoCAD definitions)

❏ Reference file clip masks

DESIGN HISTORY

You can use the Design History functionality while working in the MicroStation environment and DGN project files. However, all design history will be removed during the Save As DWG process. That should not limit your ability to take advantage of this tool internally while using MicroStation and the DWG file format.

CUSTOMIZED RESOURCES

You can use custom line styles and custom patterns. However, you must choose how you want them to save to DWG. Your first option is to drop them during the Save As DWG process. Your second option is to create matching custom line styles and/or patterns in the DWG environment so that you can remap to them during the Save As DWG process.

The use of splines in patterns cannot be duplicated in AutoCAD and should be avoided. All other custom patterns must be created for both the DGN and DWG file formats for remapping to be successful.

REFERENCE MASKS

You should not use reference file masking capabilities in MicroStation because they will be removed during the Save As DWG process.

CHOOSING THE RIGHT WORKFLOWS

Which of the previous workflows fit your project needs? Once you have determined which workflow applies to you, use the following list of questions to narrow down the more specific project requirements.

❑ What file format will the final project deliverables need to be? DWG, DGN, or both?

❑ What percentage of the files used in the project will be DWG? DGN?

❑ Who will be responsible for doing most of the editing and assembling of files? MicroStation users or AutoCAD users?

❑ Are there downstream or add-on applications requiring a particular file format? Third-party PDM software, presentation software, discipline-specific application software?

❑ Are there advanced functions of MicroStation that will facilitate the completion of the project (such as custom line styles, design history, self references, and multiple models)?

❑ Will the project files go through revisions and return to MicroStation repeatedly for reviews and changes?

The answers to these questions should help you to clarify the scenario into which your particular project falls. Specifically, the format of the deliverables and the cycling of revisions are of importance.

A: MicroStation Key-in Shortcuts

Key-in Shortcuts

AA = active angle
AC = active cell
AD = data point – delta ACS
AE = define active entity
AM = activate menu
AP = active pattern cell
AR = active cell, place relative
AS = active scale factors
AT = activate tutorial
AX = data point – absolute ACS
AZ = Set Active Depth
CC = create cell
CD = delete cell from cell library
CM = Place Active Cell Matrix
CO = active color
CR = rename cell
CT = attach color table
DA = active displayable attribute type
DB = attach control file to design file
DD = Set Display Depth (relative)
DF = open Fonts settings box
DI = data point – distance, direction
DL = data point – delta coordinates
DP = Set Display Depth
DR = display text file
DS = fence filter
DV = delete saved view
DX = data point – delta view coordinate
DZ = Set Active Depth (relative)
EL = create element list file
FF = file fence - copy fence contents to file

FI = database row as active entity
FT = active font
GO = Global Origin
GR = grid reference spacing
GU = grid unit spacing
KY = keypoint snap divisor
LC = active line style
LD = dimension level
LL = active text line length
LS = active line spacing
LT = active line terminator
LV = active level
NN = active text node number
OF = level display off
ON = level display on
OX = retrieve user command index
PA = active pattern angle
PD = active pattern spacing
PS = active pattern scale
PT = active point
PX = delete ACS
RA = attribute review selection criteria
RC = attach cell library
RD = open design file
RF = Attach Reference File
RS = name report table
RV = Rotate View (relative)
RX = Select ACS
SD = active stream delta
SF = separate fence - move fence contents to design file
ST = active stream tolerance
SV = save named view
SX = save ACS
TB = tab spacing for importing text
TH = active text height

TI = copy and increment value
TV = dimension tolerance limits
TW = active text width
TX = active text size (height & width)
UC = active user command
UCC = compile user command
UCI = active user command by index number
UR = unit distance
VI = attach named view
WO = Window Origin
WT = active line weight
XD = open design file, keep view config
XS = active X scale
XY = data point – absolute coordinates
YS = active Y scale
ZS = active Z scale

$ key-in
/   key-in
*   key-in
| key-in
;   key-in

Command Line Switches

-o   does not open any references
-m   specifies model to open
-r   opens design file in read-only
-I   passes parameters to INITAPPS
-wu   specifies user configuration
-wp   specifies project configuration

313

-wi   specifies interface configuration

-wd   specifies database configuration

-wa   specifies MDLapp to start in MS_INITAPP

-wc   specifies config to use on startup

-wr   specifies _USTN_WORKSPACEROOT

-ws   specifies config variable to define

-s   dumps text to specified startup file

-debug
-debug =1 least detail
-debug =5 most detail

-help or -? Displays command line help for *ustation.exe*

Action Types

E, keyinCommand entry keyin
T, keyinTerminated keyin
K, keyinNon-terminated keyin
M, messageMessage
C, cellnameAbsolute cell placement
R, cellnameRelative cell placement

/d   pause for data point
/k   pause for key-in
%d   pause for data point no prompts
%k   pause for key-in no prompts
null   executes NULL to suspend all commands

AccuDraw Shortcuts

Enter   Smart Lock
Space   Change Mode
O   Set Origin

V   View Rotation
T   Top Rotation
F   Front Rotation
S   Side Rotation
B   Base Rotation
E   Cycle Rotation
X   Lock X
Y   Lock Y
Z   Lock Z
D   Lock Distance
A   Lock Angle
L   Lock Index
RQ   Rotate Quick
RA   Rotate ACS
RX   Rotate about X
RY   Rotate about Y
RZ   Rotate about Z
?   Show Shortcuts
H   HOLD AccuDraw
~   Bump Tool Setti
GT   Go to Tool Sett
GK   Go to Keyin
GS   Go to Settings
GA   Get ACS
WA   Write to ACS
P   Point Keyin (si
M   Point Keyin (mu
I   Intersect Snap
N   Nearest Snap
C   Center Snap
K   Snap Divisor
U   Suspend AccuSna
J   Toggle AccuSnap
Q   Quit AccuDraw

Helpful Key-ins

Create symbol
Delete symbol
Accusnap suspend
Accusnap toggle
set reflleveloverrides on

Dialog toolbox NAME toggle
Dialog toolsettings toggle
Dialog cellmaintenance toggle
Dialog reference toggle

Leveldisplay dialog toggle
Levelmanager dialog toggle
Model dialog toggle
Raster dialog toggle
Popset toggle
View toggle
Lock axis toggle

Dialog cellmaintenance popup
Dialog cmdbrowse popup
Leveldisplay dialog popup
Levelmanager dialog popup

Dialog Palette 40line

Choose all
Choose none
Choose last

DWG audit fix
DWG recover

Change direction

Mark
Undo mark

Match color fromcursor
Match level fromcursor
Match style fromcursor
Match weight fromcursor
Match symbology fromcursor
Match element fromcursor

Mdl silentload MDLAPP

Model set annotationscale SCALE_VALUE
annotationscale add
annotationscale change
annotationscale remove

Reference adjustcolors VALUE, SATURATION, REFERENCE

Set hilite COLOR

B: AutoCAD Commands and Their MicroStation Equivalents

	AutoCAD	MicroStation Equiv	Alias	Menu or Shortcut
	+customize	…	…	Workspace > Customize
	'+dsettings	…	…	…
	+options	Preferences	…	Workspace > Preferences
	+publish	Print, Batch Print	…	File > Print, File > Batch Print
	+ucsman	Auxiliary Coordinates	…	Tools > Auxiliary Coordinates
	+view	Saved Views	…	Utilities > Saved Views
	+vports	View Groups	…	Window > Views > Dialog
A	About	Dialog Aboutustn	…	Help > About Microstation
	acadblockdialog	Dialog Cellmaintenance	Di Ce	AC=*Cellname*
	acadwblockdialog	File Fence	…	FF=
	acisin	Import ACIS	…	File > Import > ACIS SAT
	acisout	Export ACIS	…	File > Export > ACIS SAT
	adcclose	…	…	…
	adcustomnavigate	…	…	…
	adcenter	…	…	…
	adcnavigate	…	…	…
	Ai-Box	Place Slab	Pla Sl	…
	Ai-Cone	Place Cone	Pla Con	…
	ai_dim_textabove	Dimstyle Text	Di Dimset	Element > Dimension Styles > Text
	ai_dim_textcenter	Dimstyle Text	Di Dimset	Element > Dimension Styles > Text
	ai_dim_texthome	…	…	…
	ai_dish	…	…	…
	ai_dome	…	…	…
	Ai-Molc	Match Level	Mat L	…
	ai_pspace	Model Active *name*	Mo A	Viewgroups
	ai_pyramid	…	…	…
	ai_selall	Select All Elements	Ctrl + A	Edit > Select All
	Ai_Sphere	Place Sphere	Pla Sp	…
	Ai-Torus	Place Torus	Pla To	…
	Ai-Wedge	Place Wedge	Pla We	…
	aidimfliparrow	…	…	…
	aidimprec	Dimstyle Units	Di Dimset	Element > Dimension Styles > Units
	aidimstyle	Dimension Styles	Di Dimset	Element > Dimension Styles
	aidimtextmove	Dimstyle Leader	Di Dimset	Element > Dimension Styles > Dimension with Leader
	aliasedit	Accudraw Shortcuts	…	Accudraw Dialog Shortcuts | Edit
	Align	Align Element	…	…
	alignspace	…	…	…
	ameconvert	…	…	…
	Aperture	Locate Tolerance	…	Tools > Preferences > Operation > Locate Tolerance

	AutoCAD	MicroStation Equiv	Alias	Menu or Shortcut
	Appload	Mdl Load	Mdl L	…
	Arc	Place Arc	Pla A	…
	Archive	Archive	…	Utilites > Archive
	arctext	Place Text Along	Pl Tex Al	…
	Area	Measure Area	Mea Ar	…
	Array	Array	Ar	…
	Arx	Mdl Load	…	Utilities > Mdl Applications
	aseadmin	Database	…	Settings > Database
	aseexport	Database	…	Settings > Database
	aselinks	Database	…	Settings > Database
	aserows	Database	…	Settings > Database
	aseselect	Database	…	Settings > Database
	asesqled	Database	…	Settings > Database
	assist	Tracking	…	Help > Tracking
	assistclose	…	…	…
	attachurl	Engineering Links	…	Tools > Engineering Links
	Attdef	Mdl L Tags Define	…	Element > Tags > Define
	Attdisp	Tags Display	…	Settings > View Attributes > Tags
	Attedit	Edit Tags	Edi T	…
	Attext	Mdl L Tags Report	…	Element > Tags > Generate Reports
	attin	…	…	…
	attout	…	…	…
	attredef	Define Tags	…	Element > Tags > Define
	attsync	…	…	…
	Audit	Mdl Load Fixrange	Mdl L Fixrange	…
B	Background	Active Background	Act Ba	…
	baction	…	…	…
	bactionset	…	…	…
	bactiontool	…	…	…
	Base	Active Origin	GO=	…
	bassociate	…	…	…
	battman	Edit Tags	Edi T	…
	battorder	…	…	…
	bauthorpalette	…	…	…
	bauthorpaletteclose	…	…	…
	bcount	…	…	…
	bcycleorder	…	…	…
	bedit	…	…	…
	bextend	Extend	Ext	…
	bgripset	…	…	…
	Bhatch	Pattern	Pat	…
	blipmode	…	…	…
	Block	Define Cell	D C	CC=*Cellname*
	block?	…	…	…
	blockicon	…	…	…
	blockreplace	Replace Cells	Rep C	…
	blocktoxref	…	…	…
	blookuptable	…	…	…

	AutoCAD	MicroStation Equiv	Alias	Menu or Shortcut
	Bmake	Dialog Cellmaintenance	Di Ce	CC=*Cellname*
	bmod	Create Cell	Cre C	CC=*Cellname*
	bmpout	Screen Capture	Ca D	Utilities > Image > Capture
	bonusmenu	…	…	…
	bonuspopup	…	…	…
	borrowlicense	…	…	…
	Boundary	Create Shape Automatic	Cr S A	…
	Boundary or Bpoly	Create Shape Automatic	Cr S A	…
	Box	Place Slab	Pla Sl	…
	bparameter	…	…	…
	bpoly	Create Region	Cre R	…
	break	Delete Partial	Del Pa	…
express	breakline	…	…	…
	browser	Browser Connect	Br A *name*	…
	browser2	…	…	…
	bsave	…	…	…
	bsaveas	…	…	…
	bscale	…	…	…
	btrim	…	…	…
	burst	Drop Element	Dro	…
	bvhide	…	…	…
	bvshow	…	…	…
	bvstate	…	…	Utilities > Microstation Link
C	Cal	Accudraw Activate	A A	…
	camera	Camera	ca	…
	cdorder	…		…
	Chamfer	Chamfer	Ch	…
	Change	Change Element Properties	Ch E E	…
	checkstandards	…	St e	Utilites > Standards Checker > Check
	chk	Check Standards	St e	
	Chprop	Change Element Properties	Ch E E	…
	chspace	…	…	…
	churls	Engineering Links	…	…
	Circle	Place Circle	Pla Ci	…
	cleanscreenoff	…	…	…
	cleanscreenon	…	…	…
	clipit	Reference Clip, Clip Volume	…	…
	close	Close	Clo	…
	closeall	…		…
	Color or Colour	Active Color	CO=	CO=*color or number*
	commandline	Key in Browser	Di Cm	…
	commandlinehide	…	…	…
	Compile	…	UCC=	…
	Cone	Place Cone	Pl Con	…
	config	Workspace Configuration	…	Workspaces > Configuration

	AutoCAD	MicroStation Equiv	Alias	Menu or Shortcut
	content	...	...	...
	Convert	...	...	...
	convertctb	...	...	...
	convertplines	...	...	...
	convertpoly	...	...	...
	convertpstyles	...	...	...
	Copy	Copy	Cop	Shift + F5
	copybase	...	...	...
	Copyclip	Clipboard Copy	Cli C	Ctrl + C
	copyhist	Message Center	...	...
	Copylink	Capture View Contents	...	Utilities > Image > Capture
	copym	Copy	Cop	...
	copytolayer	...	...	...
	count	...	...	...
	cui	Workspace Customize	...	Workspace > Customize
	cuiexport	...	...	...
	cuiimport	...	...	...
	cuiload	...	...	...
	cuiunload	...	...	...
	customize	Workspace Customize	...	Workspace > Customize
	Cutclip	Clipboard Cut	Cli Cu	Ctrl + x
	Cylinder	Place Cylinder	Pla Cy	...
D	date	...	...	...
	dbclose	Database	...	Settings > Database
	dbconnect	Database	...	Settings > Database
	dblkclkedit	Database	...	Settings > Database
	dblist	Database	...	Settings > Database
	dbtrans	Database	...	Settings > Database
	Ddattdef	Edit Tags	Edi T	...
	Ddatte	Edit Tags	Edi T	...
	Ddattext	Mdl Load Tags Report	...	...
	Ddchprop	Change Element Properties	Ch E E	...
	Ddcolor	Active Font	FT=	FT=*font number or name*
	Ddedit	Edit Text	Edi Te	...
	Ddgrips	Handles	...	Workspaces > Preferences > Input > Highlight Selected
	Ddim	Dialog Dimsettings Open	...	Element > Dimensions
	Ddinsert	Dialog Cellmaintenance	Di Ce	AC=*Cellname*
	ddlmodes	Level Manager	Levelm D O	...
	ddltype	Line Styles Custom	Lines S	Element > Line Styles > Custom
	Ddmodify	Element Information	An	Ctrl + I
	ddosnap	Active Snaps, Multi-Snaps	...	Settings > Snaps
	ddplotstamp	Plot Stamp, Text Substitution	...	Printing only
	Ddptype	Active Scale	...	AS=*scale factor*

	AutoCAD	MicroStation Equiv	Alias	Menu or Shortcut
	Ddrmodes	Mdl Load Dgnset	Mdl L Dgnset	Settings > Design File
	Ddselect	Select	Se	Tool Settings: Method/Mode
	Dducs	Dialog Coordsys	Di Coo	…
	Dducsp	Dialog Coordsys	Di Coo	…
	Ddunits	Dialog Units	Di U	…
	Ddview	Dialog Namedviews	Di Namedv	SV=*ViewName*, VI=*ViewName*, DV=*ViewName Save*, Restore, Delete Views
	Ddvpoint	Dialog Rotateucs	Di Ro	…
	Delay	Pause	Pau	…
	detachurl	Engineering Links	…	…
Dimensions				
	Dim	Dimension	Dim	…
	DimAligned	Dimension Element	Dim E	Tool Settings: Alignment>True
	DimAngular	Dimension Angle Lines	Dim A L	…
	DimBaseline	Dimension Linear	Dim L	Tool Settings: Dimension Stacked On
	DimCenter	Dimension Center Mark	Dim C M	…
	DimContinue	Dimension Linear	Dim L	Tool Settings: Dimension Stacked Off
	DimDiameter	Dimension Diameter	Dim D	…
	dimdisassociate	Dimension Update	Dim Up	…
	dimedit	Dimension Element	Dim E	…
	dimex	…	…	…
	dimim	Dimension Style Import	…	Element > Dimension Styles > Import
	dimjogged	…	…	…
	DimLeader	Place Note	Pla Not	Optional: Place Note Multi
	Dim:Ordinate	Dimension Ordinate	Dim 0	…
	Dim:Override	Dimension Update	Dim Up	…
	Dim:Radius	Dimension Radius	…	…
	dimreassociate	Dimension Reassociate	Dim Re	…
	dimregen	Update View	Up	F9
		Dimension Update	Dim Up	…
	DimStyle	Dimension Style	Dims	Element > Dimension Styles
	DimTedit	Modify Element	Modi	…
	Diml	Dimension	Dim	…
	Dimaligned	Dimension Element	Dim E	Tool Settings: Alignment > True
	Dimangular	Dimension Angle Lines	Dim A L	…
	Dimbaseline	Dimension Linear	Dim L	Tool Settings: Dimension Stacked On
	Dimcenter	Dimension Center Mark	Dim C M	…
	Dimcontinue	Dimension Linear	Dim L	Tool Settings: Dimension Stacked Off
	Dimdiameter	Dimension Diameter	Dim D	…
	Dimedit	Modify Element	Modi	…
	Dimlinear	Dimension Linear	Dim L	…
	Dimordinate	Dimension Ordinate	Dim 0	…
	Dimoverride	Dimension Update	Dim Up	…
	Dimradius	Dimension Radius	…	…
	Dimstyle	Dimension Styles	Setm S D	Element > Dimension Styles
	Dimtedit	Modify Element	Modi	…
	Dist	Measure Distance Points	Meas	…
	Divide	Construct Point Between	…	…

	AutoCAD	MicroStation Equiv	Alias	Menu or Shortcut
	dline	Place Multi-Line	…	…
	Doughnut Or Donut	Place Circle	Pla Ci	Tool Settings: Fill Type > Opaque
	Dragmode	Set Dynamic On/Off/ Toggle	…	Settings > View Attributes > Dynamic
	drawingrecovery	…	…	…
	drawingrecoveryhide	…	…	…
	Draworder	Wset Add and Wset Drop	…	…
	dropgeom	Clear Selection	…	…
	dsettings	Design File Settings	…	Settings > Design File > Grid
	Dsviewer	…	…	Window > Open/Close
	Dtext	Place Text	Pla Tex	…
	Dview	Rotate View Extended	…	…
	Dwfout	…	…	…
	dwfoutd	…	…	…
	dwglog	…	…	…
	dwgprops	…	…	…
	dxbin	…	…	…
	dxbout	…	…	…
	Dxfin	Dxf In	Dx I	File > Import > Dwg Or Dxf
	Dxfout	Dxf Out	Dx 0	File > Export > Dwg Or Dxf
E	eattedit	Edit Tags	Edi T	…
	eattext	Generate Reports	…	Element > Generate Reports
	edge	…		…
	edgesurf	Construct Surface by Edge	…	…
	edittime	…	…	…
	Elev	Active Zdepth Absolute	Act Z A	AZ=
	Elevation	Set Active Depth	Act Z A	AZ=
	Ellipse	Place Ellipse	Pla E	…
	End (Not In R14)	Exit	Exi	File > Exit
	eplotext	…	…	…
	Erase	Delete	Del	Alt + F5
	etransmit	Packager	…	Utilities > Packager
	exoffset	Copy Parallel	Cop P	…
	exp	Save Image As	…	Utilities > Image > Save
	explan	Rotate View Extended	Ro V E	VI=TOP
	Explode	Drop Element	Dro	…
	Export	…	Dw O or Dx O	File > Export > Dwg or Dxf
	expressmenu	…	…	…
	expresstools	…	…	…
	Extend	Extend Line Intersection	Ext L I	…
	extrim	Trim, Intelli-trim	Tri	…
	Extrude (Acis)	Extrude Surface Region	Extru S R	…
F	field	…	…	…
	FileOpen	Dialog Openfile	Di 0	Rd=
	Fill	Set Fill On/Off	…	Ctrl + B > Fill
	Fillet	Fillet	Fill M	…
	Filter	…	…	Edit > Select By Attributes

	AutoCAD	MicroStation Equiv	Alias	Menu or Shortcut
	find	Find and Replace	...	Edit > Find and Replace
	finish	...	...	Settings > Rendering > Materials
	flatten	Place Cell	Pl Ce	Tool Settings: Flatten
	Fog	Dialog Viewrenderset	Di Viewr	Settings > Rendering > View Attributes
	fs	...	...	...
	fsmode	...	...	...
	fullscreen	...	...	...
	fullscreenoptions	...	...	...
G	gatte	Change Tags	Chan T	Setting: All
	getsel	Select By Attributes	...	Edit > Select By Attributes
	gotourl	Follow Engineering Link	...	Tools > Engineering Links
	gradient	...	...	...
	Grid	Active Gridunit	GR=	Settings > Design File > Grid
	grips	Handles	...	
	Group	Group Selection	Gr S	Ctrl + G
H	Hatch	Pattern	Pat	...
	hatchedit	Change Pattern	Pat Modi	...
	Help or '?	Help	He	F1
	Hide	Render View Hidden	Rend V H	...
	hlsettings	Export Visible Edges	...	File > Export > Visible Edges
	hyperlink	Engineering Links	...	Tools > Engineering Links
	hyperlinkbase	...	...	...
	hyperlinkfwd	...	...	...
	hyperlinkopen	Follow Engineering Link	...	Tools > Engineering Links
	hyperlinkoptions	...	...	...
	hyperlinkstop	...	...	...
I	Id	...	...	*Tentative Point Snap to Display XYX Coordinate*
	igesin	Import IGES	...	File > Import > IGES
	igesout	Export IGES	...	File > Export > IGES
	Image	Raster Attach Interactive	Ras A I	...
	Imageadjust	Mdl L Imagevue	...	Image > Gamma Correction
	Imageattach	Raster Attach Interactive	Ras A I	...
	Imageclip	Raster Clip Boundary	Ras C B	...
	imageedit	...	...	...
	imageframe	...	...	...
	Imagequality	Mdl L Imagevue	...	...
	Import	...	Dw I or Dx I	File > Import > Dwg Or Dxf
	Insert	Dialog Cellmaintenance	Di Ce	Ac=Cellname
	Insertobj	...	...	Edit > Insert Object
	inserturl	Attach Engineering Link	...	Tools > Engineering Links
	interfere	Intersection Feature	...	Tools > Feature Modeling > Intersection
	intersect	Intersection Feature	...	Tools > Feature Modeling > Intersection
	isoplane	...	...	...
J	join	Create Complex Shape	Cre S	Tool Settings: Simplify Geometry

	AutoCAD	MicroStation Equiv	Alias	Menu or Shortcut
	jpgout	Export Image	…	Utilities > Image
	justifytext	…	…	…
L	laycur	Match Element Attributes	Mat	Tool Settings: Level
	laydel	Level Purge	Le P	…
	Layer	Active Level	Act L	LV=
	layerp	…	…	…
	layerpmode	View Previous	Vi P	…
	layfrz	LevelDisplay Global-Freeze	…	Level Display > Global Freeze
	layiso	All Except Element	…	Level Display > All Except Element
	laylck	…	…	…
	laymch	…	…	…
	laymgr	Level Manager	…	…
	layoff	Off By Element	…	Level Display > Off by Element
	layon	All Levels On	…	Level Display > Levels On
	layout	Create Model	…	Models > Sheet Models
	layoutmerge	…	…	…
	layoutwizard	…	…	…
	laythw	All Levels On	…	Level Display > Levels On
	laytrans	Remapping Utilities	…	File > Save As > Options
	layulk	…	…	
	layvpi	All Except Element	…	Level Display > All Except Element
	layvpmode	…	…	…
	laywalk	…	LV=+1; OF=ALL	…
	Leader	Place Note	Pla Not	…
	Lengthen	Extend Line	Ext L	…
	Light	Light Define	Li D	…
	Limits	…	…	Settings > Design File
	Line	Place Line	Pla L	…
	Linetype	Active Terminator	…	LC=
	Linetype or Ddltype	Active Style #	Act S #	LC=
	List	Analyze Element	An	Ctrl + I
	listurl	…	…	…
	lman	…	…	…
	Load	Mdl Load	Mdl L	Utilities > Mdl Applications
	logfileoff	…	…	…
	logfileon	…	…	…
	lsedit	…	…	…
	lslib	…	…	…
	lsnew	…	…	…
	lsp	user commands	UC=	…
	lspsurf	…	…	…
	ltscale	Active Linestylescale	Act Lines	…
	lweight	Line Weights	WT=	Workspace > Preferences > View Options
M	Markup	Redline	…	…
	markupclose	…	…	…

	AutoCAD	MicroStation Equiv	Alias	Menu or Shortcut
	Massprop	Measure Volume	Mea V	Tool Settings: Mass Properties
	matchcell	…	…	…
	Matchprop	Match Element	Mat E	…
	Matlib	Material Palette Open	Mate P O	…
	Measure	Construct Point	…	…
	meetnow	…	…	…
	Menu	Attach Menu	AM=	Workspace > Customize
	Menuload	Attach Menu	AM=	…
	Menuunload	Detach Menu	AM=	Workspace > Customize
	Minsert	Place Cell Matrix	CM=	CM=(Matrix Cell)
	Mirror	Mirror Original	Mi O	Tool Settings:
	Mirror3d	Mirror Original	Mi O	Tool Settings:
	mkltype	Create Custom Line Style	…	Element > Line Styles > Edit
	mkshape	Create Symbol	…	…
	Mledit	Dialog Toolbox Joints	Di T J	…
	Mline	Place Mline	Pla Ml	…
	Mlstyle	Dialog Multiline Open	Dial Mu O	…
	mocoro	…	…	…
	model	…	…	…
	Move	Move	Mov	Ctrl + F5
	movebak	MS_BACKUP	…	Workspace > Configuration
	mpedit	All edits are multiple	…	…
	mredo	All Redo are multiple	…	…
	Mslide	…	…	Utilities > Render > Animation
	mspace	…	…	…
	mstretch	Fence Stretch	…	…
	Mtedit	Edit Text	Edi Te	…
	Mtext	Place Dialogtext	Pla Di	…
	Mtprop	Edit Text	Edi Te	…
	multiple	All commands are multiple		…
	mview	Clip Reference	Ref C	Reference File dialog
	mvsetup	Copy/Fold Reference	Ref C F	Reference File dialog
N	ncopy	Copy from Reference	…	Settings: Locate On
	netload	…	…	
	New	Create Drawing	Cr D	Ctrl + N
	newsheetset	…	…	…
O	Offset	Copy Parallel	Cop P	…
	oldmtext	Place Text	Pl Tex	…
	oldmtprop	…	…	…
	Olelinks	Edit Links	…	Edit > Dde Links
	oleopen	Open Links	…	Edit > Links
	olescale	…	…	…
	Oops	Undo	…	Ctrl + Z
	Open	Newfile	RD=	Ctrl + O
	opendwfmarkup	…	…	…
	OpenSheetSet	…	…	…
	Openurl	Engineering Links	…	File > Open Url

	AutoCAD	MicroStation Equiv	Alias	Menu or Shortcut
	options	Preferences, Configuration	…	Workspace > Preferences or Configuration
	Ortho	Lock Axis	…	…
	Osnap or Ddosnap	…	…	Tentative Snap
	overkill	Data Cleanup	…	Utilities > Data Cleanup
P	pagesetup	Plot Configurations	…	…
	painter	Match Tools	Mat	Tools > Match
	-Pan	Pan View	Pan	View > Scroll Bar Buttons
	Pan or Rtpan	Dynamic Pan	…	Shift + Datapoint
	partialload	…	…	
	pasteashyperlink	…	…	Edit > DDE Links
	pasteblock	…	…	
	Pasteclip	…	…	Edit > Paste
	pasteorig	…	…	
	Pastespec	…	…	Edit > Paste Special
	pcinwizard	…	…	…
	pcxin	…	…	…
	Pedit	Modify Element	Modi E	…
	Pedit	Insert Vertex	Ins V	…
	Pedit	Delete Vertex	Del V	…
	pface	Create Planar Surface	…	Tools > Surface Modeling
	plan	Rotate View	VI=TOP	…
	Pline	Place Smartline	Pl Sm	…
	pljoin	Create Chain	…	…
	Plot	Plot	Print	Ctrl + P
	plotstamp	Plot Attributes, Text Substitution	…	…
	plotstyle	.PLT plot driver file	…	…
	plottermanager	…	…	…
	plt2dwg	…	…	…
	pngout	Save Image	…	Utilities > Image
	Point	Place Point	Pl Po	…
	Polygon	Place Polygon	Pl Pol	…
	pqcheck	…	…	…
	Preferences	…	…	Workspace > Preferences
	Preview	Preview	Prev	…
	properties	Attributes, Element Information	…	…
	propertiesclose	…	…	…
	propulate	…	…	…
	psbscale	…	…	…
	psdrag	…	…	…
	psetupin	…	…	…
	psfill	…	…	…
	psin	Paste Image	…	Edit > Paste, Ctrl + V
	psout	Save Image	…	Utilites > Image
	pspace	…	…	…
	psltscale	…	…	…
	publish	Print to PDF	…	File > Print, Ctrl + P

	AutoCAD	MicroStation Equiv	Alias	Menu or Shortcut
	publsihtoweb	…	…	…
	purge	Compress Design	Com D	Shift + F3
Q	qcclose	…	…	…
	qdim	Dimension Multiple Elements	…	Tool Setting: Select Multiple Elements
	qlattach	Reassociate Note	…	…
	qlattachset	Reassociate Note	…	…
	qdetachset	Drop Association	Dro A	…
	qleader	Place Note	Pl Not	…
	qnew	…	…	…
	qquit	Save Design	S D	Ctrl + S
	qsave	Save	S D	File > Save, Ctrl + S
	qselect	PowerSelector, Select By Attributes	…	Edit > Select By Attributes
	Qtext	Fast Font	…	Ctrl + A
	quickcalc	…	…	…
	Quit	Quit	Q	…
R	r14penwizard	…	…	…
	ray	…	…	…
	Recover	Mdl L Fixrange	…	…
	Rectangle	Place Block	Pl B	…
	Redefine	…	…	Workspace > Configuration
	redir	…	…	Workspace > Configuration
	redirmod	…	…	Workspace > Configuration
	Redo	Redo	…	Ctrl + R
	Redraw	Update View	Up V	F9
	Redrawall	Update All	Up A	…
	refclose	…	…	…
	refedit	Exchange Reference	…	Reference dialog > Exchange
	refset	…	…	…
	regen	Update View	Up V	…
	regenall	Update View	Up V	…
	regenauto	Update View	Up V	…
	Region (Acis)	Group Holes	Gr H	…
	reinit	…	…	…
	rename	…	…	…
	Render	Render View	Rend V	…
	renderundload	…	…	…
	renderupdate	…	…	…
	rendscr	…	…	…
	Replay	…	…	Utilities > Image > Animation
	repurls	…	…	…
	resetblock	…	…	…
	resume	…	…	…
	returnlicense	…	…	…
	revcloud	…	…	…
	revert	…	…	…
	revolve	Construct Revolution	…	…
	revsurf	Construct Revolution	…	Tool Setting: Surface

	AutoCAD	MicroStation Equiv	Alias	Menu or Shortcut
	rfileopt	…	…	…
	Rmat	Material Palette Open	Mate P O	…
	Rotate	Rotate	Ro	…
	rotate3d	Rotate	…	…
	Rpref	…	…	Settings > Rendering > Setup
	Rscript	…	…	Utilities > Macros or Utilities > Run
	rtedit	Text Substitution	…	…
	rtext	Text Substitution	…	…
	rtpan	Pan	…	Shift + Data drag
	rtucs	Rotate ACS	…	…
	rtzoom	Wheel Zoom	…	Roll Wheel Mouse
	rulesurf	…	…	…
S	Save	Save Design	…	…
	saveall	…	…	…
	Saveas	Dialog Saveas	Di S	File > Save As…
	saveasr12	Save as DWG		File > Save As…
C	Saveimg	Dialog Saveimage	Di Savei	Utilities > Image > Save
	Saveurl	…	…	Utilities > Microstation Links
	Scale	Scale	Sc	…
	scalelistedit	Edit file Scales.def	…	…
	scaletext	Annotation Scale	…	…
	Scene	…	…	Utilities > Image > Animation
	Script	Macros	…	Utilities > Macros
	Section	…	…	Utilities > Generate Section
	securityoptions	…	…	
	Select	Choose Element	…	F5
	Selecturl	…	…	Utilities > Microstation Links
	setidrophandler	…	…	
	send	…	…	File > Send
	Setuv	…	…	Settings > Rendering > Assign Materials
	Setvar	Set	Set	…
	Shade	Render View Filled	Rend V F	Settings > Rendering > Define Materials
	shademode	Render View	Rend V	Settings > Rendering > View Attributes
	shape	Place Symbol	Pl Sy	…
	sheetset	…	…	…
	sheetsethide	…	…	…
	Shell or Sh	!	!	…
	showmat	…	…	…
	showurls	Show Engineering Links	…	…
	shp2blk	…	…	…
	sigvalidate	…	…	…
	Sketch	Place Lstring Stream	Pl Ls St	…
	slice	…	…	…
	Snap	Lock Grid	…	…
	soldraw	Export Hidden Lines	…	File > Export > Hidden Lines
	solid	Place Solid	…	…
	solidedit	Modify Solid	…	…
	solprof	…	…	…
	solview	Render View	Ren V	Utilities > Render > Hidden Line

	AutoCAD	MicroStation Equiv	Alias	Menu or Shortcut
	spacetrans	…	…	…
	spell	Spell Checker	…	…
	Sphere	Place Sphere	Pl Sp	…
	Spline	Place Bspline	Pl Bs	…
	Splinedit	Modify Bspline Curve	Mod B C	…
	ssx	…	…	…
	standards	Standards Checker	…	Utilities > Standards Checker
	stats	…	…	…
	status	…	…	…
	Stretch	Fence Stretch	F St	…
	Style or Ddstyle	Active Font	FT=Font#	Element > Text
	stylesmanager	Line Style Editor	…	Element > Line Styles > Edit, Custom
	subtract	Construct Difference	…	…
	superhatch	…	…	…
	sysvdlg	…	…	…
	syswindows	Window Tile	w T	Window > Tile, Arrange
T	Table	…	…	…
	tabledit	…	…	…
	tablexport	…	…	…
	tablestyle	…	…	…
	Tablet	Digitizer Setup	…	…
	tabsurf	Extrude along Path	…	…
	taskbar	…	…	…
	tbconfig	…	…	…
	tcase	…	…	…
	tcircle	…	…	…
	tcount	Increment Text	Incr T	…
	Text	Place Text	Pl Tex	…
	textfit	Place Text Fitted	Pl Tex F	Tool Settings: Fitted
	textmask	Text Style Background	…	Element > Text Styles
	textscr	Message Center	…	Status Bar
	texttofront	…	…	…
	textunmask	Text Style Background	…	Element > Text Styles
	tframes	…	…	…
	tiffin	Import Image	…	File > Import > Image
	tifout	Save Image	…	Utilities > Image
	time	…	…	…
	tinsert	…	…	…
	tjust	…	…	…
	today	…	…	…
	Tolerance	Mdl L Geomtol	…	…
	Toolbar	Tool Boxes	…	Tools > Tool Boxes or Ctrl + T
	toolbox	Toolbar, Toolframes	…	…
	toolpalettes	…	…	…
	toolpalettesclose	…	…	…
	torient	…	…	…
	Torus	Place Torus	Pl To	…
	trace	…	…	…
	transparency	…	…	…

	AutoCAD	MicroStation Equiv	Alias	Menu or Shortcut
	traysettings	…	…	…
	treestat	…	…	…
	Trim	Trim, Intellitrim	Tri	…
	tscale	…	…	…
	tutclear	…	…	…
	tutdemo	…	…	…
	text2mtext	…	…	…
	txtexp	Drop Text	Dro	…
U	U	Undo	Undo E	Ctrl + Z
	Ucs	Dialog Coordsys	Di Coo	SX=*AscName*, RX=*AcsName*, PX=*Acs-Name*, Save Acs, Attach Acs, Delete Acs
	Ucsicon	Set Acsdisplay On/Off/ Toggle	…	Ctrl + B
	Undefine	…	…	Workspace > Configuration
	ucsman	Define ACS	…	
	undefine	…	…	
	Undo	Undo	Undo	Ctrl + Z
	union	Construct Union	…	…
	Units	Dialog Units	Di U	…
	updatefield	…	…	…
	updatethumbsnow	…	…	…
V	vbaide	…	…	…
	vbaload	VBA Load	…	Utilities > Macro > Project Manager
	vbaman	…	…	Utilities > Macro > Project Manager
	vbarun	VBA Run	…	Utilities > Macro > Project Manager
	vbastmt	…	…	Utilities > Macro > Project Manager
	vbaunload	VBA Unload	…	Utilities > Macro > Project Manager
	View	Dialog Namedviews	…	SV=, VI=, DV=
	viewplotdetails	…	…	…
	viewres	…	…	…
	vlide	…	…	…
	vpclip	Clip Reference	Ref C	File > Reference
	vplayer	…	…	…
	vpmax	…	…	…
	vpmin	…	…	…
	Vpoint	Rotate View	Rot V	RV=
	vports	…	…	…
	vpscale	…	…	…
	vpsync	…	…	…
	Vslide	…	…	Utilities > Image > Animation
	vtoptions	…	…	
W	Wblock	File Fence	…	FF=
	Wedge (Acis)	Place Wedge	Pl W	…
	whohas	…	…	…
	wipeout	…	…	…
	wmfin	Import Image	…	File > Import > Image
	wmfout	Save Image	…	Utilities > Image
	wmfopts	…	…	…
	workspace	Customize	…	Workspace > Customize

	AutoCAD	MicroStation Equiv	Alias	Menu or Shortcut
	wssave	…	…	…
	wssettings	…	…	…
X	Xattach	Reference Attach	Ref At	RF=*Filename*
	Xbind	Reference Detach	Ref D	…
	Xclip	Reference Clip Boundary	Ref C B	…
	xdata	…	…	…
	xdlist	…	…	…
	xline	…	…	…
	xlist	…	…	…
	xopen	Reference Exchange	…	File > Reference
	Xplode	Drop Element	Dro	Tool Settings: Controls
	Xref	Dialog Reference	…	File > Reference
	xrefclip	Clip Reference	…	File > Reference
Z	Zoom or Rtzoom	Zoom	Z	F11
?	?	? Or $	…	…
3	3D	…	…	…
	3darray	…	…	…
	3dclip	Clip volume	…	…
	3dconfig	…	…	…
	3dcorbit	Rotate View	Rot V	…
	3ddistance	…	…	…
	3ddwfpublish	…	…	…
	3dface	Create Planar Surface	…	…
	3dmesh	…	…	…
	3dorbit	Rotate View	Rot V	…
	3dorbitctr	…	…	…
	3dpan	…	…	…
	3dpantransparent	…	…	…
	3dpoly	Smartline	Pl Sm	…
	3drender	Render View	Rend V	Utilities > Render
	3dsin	…	…	…
	3dsout	…	…	…
	3dswivel	…	…	…
	3dzoom	…	…	…
	3dzoomtransparent	…	…	…

Index